Beginning Data Science with R

Manas A. Pathak

Beginning Data Science with R

Springer

Manas A. Pathak
Sunnyvale
California
USA

Additional material to this book can be downloaded from http://extras.springer.com.

ISBN 978-3-319-12065-2 ISBN 978-3-319-12066-9 (eBook)
DOI 10.1007/978-3-319-12066-9

Library of Congress Control Number: 2014958137

Springer Cham Heidelberg New York Dordrecht London

Printed on acid-free paper

Springer is part of Springer Science+Business Media (www.springer.com)

Dedicated to my wife Esha and to my parents
Dr. Ashok and Dr. Sulochana Pathak

Preface

We live in an age where data is being collected at unprecedented levels. One can benefit from data only after converting it into useful and actionable knowledge. In the past few years, the methodology of extracting insights from data or "data science" has emerged as a discipline in its own right. Historically, a diverse set of tools have been used for data analysis. The R programming language is increasingly becoming a one-stop solution to data science.

R is an open-source software and can be used on most computing platforms: Windows, Unix/Linux, Mac OS X. The primary reason for the growing popularity of R is due to a vast package library containing implementations for most statistical analysis techniques. On the other hand, R is not an easy programming language to learn due to its esoteric syntax. The documentation for most packages is in the form of reference material, which makes it difficult for programmers without a background in statistics to get started on their own.

Goal of this Book

In this book, we introduce the readers to some of the useful data science techniques and their implementation with the R programming language. We attempt to strike a balance between the how and the why. We will cover both the how: various data science processes and methodologies, and the why: understanding the motivation and internals of each technique. The goal of this book is to enable the reader to apply similar methodologies to analyze their own datasets.

We aim to provide a tutorial for performing data science with R as opposed to a reference manual. The number of data analysis methods and the corresponding R packages is vast. This book is intended to be a good starting point to begin data analysis. In each chapter, we present a data science methodology with a case study, including a dataset. All the datasets and source code listings are available for download at: http://extras.springer.com.

Prerequisites

This book is intended for the readers who are not familiar with statistical analysis. When covering each data analysis task, we will provide a high-level background of the underlying statistics. Experience with at least one programming language is a prerequisite. This book will be useful for readers who are not familiar with the R programming language. We will review the fundamentals of R in Chapter 2 along the notations and conventions used in this book.

August 2014
Sunnyvale, California

Dr. Manas A. Pathak

Contents

1 Introduction 1
1.1 What Is Data Science? 1
1.2 Why R? 2
1.3 Goal of This Book 3
1.4 Book Overview 3
References 4

2 Overview of the R Programming Language 5
2.1 Installing R 5
2.1.1 Development Tools 6
2.2 R Programming Language 7
2.2.1 Operators 7
2.2.2 Printing Values 8
2.2.3 Basic Data Types 9
2.2.4 Control Structures 10
2.2.5 Functions 12
2.3 Packages 13
2.3.1 R Help System 15
2.4 Running R Code 16
Reference 17

3 Getting Data into R 19
3.1 Reading Data 20
3.1.1 Text Files 20
3.2 Cleaning Up Data 25
3.2.1 Identifying Data Types 26
3.2.2 Data Entry Errors 27
3.2.3 Missing Values 28
3.3 Chapter Summary 30
References 30

4 Data Visualization 31
4.1 Introduction 31
4.2 Basic Visualizations 32
4.2.1 Scatterplots 33
4.2.2 Visualizing Aggregate Values with Bar plots and Pie charts 39
4.2.3 Common Plotting Tasks 45
4.3 Layered Visualizations Using ggplot2 48
4.3.1 Creating Plots Using qplot() 48
4.3.2 ggplot(): Specifying the Grammar of the Visualization 53
4.3.3 Themes 55
4.4 Interactive Visualizations Using Shiny 55
4.5 Chapter Summary and Further Reading 59
References 60

5 Exploratory Data Analysis 61
5.1 Summary Statistics 62
5.1.1 Dataset Size 62
5.1.2 Summarizing the Data 63
5.1.3 Ordering Data by a Variable 65
5.1.4 Group and Split Data by a Variable 66
5.1.5 Variable Correlation 68
5.2 Getting a Sense of Data Distribution 71
5.2.1 Box Plots 71
5.2.2 Histograms 75
5.2.3 Measuring Data Symmetry Using Skewness and Kurtosis 80
5.3 Putting It All Together: Outlier Detection 82
5.4 Chapter Summary 84
References 85

6 Regression 87
6.1 Introduction 87
6.1.1 Regression Models 88
6.2 Parametric Regression Models 89
6.2.1 Simple Linear Regression 90
6.2.2 Multivariate Linear Regression 99
6.2.3 Log-Linear Regression Models 101
6.3 Nonparametric Regression Models 103
6.3.1 Locally Weighted Regression 104
6.3.2 Kernel Regression 107
6.3.3 Regression Trees 109
6.4 Chapter Summary 114
References 114

7 Classification 115
7.1 Introduction 115
7.1.1 Training and Test Datasets 116
7.2 Parametric Classification Models 117
7.2.1 Naive Bayes 117
7.2.2 Logistic Regression 122
7.2.3 Support Vector Machines 126
7.3 Nonparametric Classification Models 130
7.3.1 Nearest Neighbors 131
7.3.2 Decision Trees 133
7.4 Chapter Summary 135
References 136

8 Text Mining 137
8.1 Introduction 137
8.2 Dataset 138
8.3 Reading Text Input Data 138
8.4 Common Text Preprocessing Tasks 141
8.4.1 Stop Word Removal 142
8.4.2 Stemming 143
8.5 Term Document Matrix 144
8.5.1 TF-IDF Weighting Function 147
8.6 Text Mining Applications 149
8.6.1 Frequency Analysis 149
8.6.2 Text Classification 151
8.7 Chapter Summary 157

Chapter 1
Introduction

1.1 What Is Data Science?

We live in the age of data. In the present day, data is all around us and collected at unprecedented levels. The data can be in the form of network/graph data: a wealth of information in a billion user social network, web pages indexed by a search engine, shopping transactions of an e-commerce business, or a large wireless sensor network. The amount of data that we generate is enormous: in 2012, every day, we created 2.5 quintillion bytes or 2.5 million terabytes of data. The growth rate is even more staggering: 90% of world's data was generated over the last two years [1].

Data is not very useful by itself unless it is converted into knowledge. This knowledge is in the form of insights, which can provide a lot of information about the underlying process. Corporations are increasingly becoming more *data driven*: using insights from the data to drive their business decisions. A new class of applications is the data product [2], which takes a step further by converting data insight into a usable consumer product.

Some of the prominent examples of data products include:

- *Google flu trends*: By analyzing the search engine query logs, Google is able to track the prevalence of influenza faster than the Centers for Disease Control and Prevention (CDC).
- *Netflix recommendation engine*: Looking at the movie ratings and watching patterns of pairs of users, the Netflix recommendation engine is able to accurately predict the ratings for the movies that a user has not seen before.

The methodology of extracting insights from data is called as *data science*. Historically, data science has been known by different names: in the early days, it was known simply as *statistics*, after which it became known as *data analytics*. There is an important difference between data science as compared to statistics and data analytics. Data science is a multi-disciplinary subject: it is a combination of statistical analysis, programming, and domain expertise [3]. Each of these aspects is important:

M. A. Pathak, *Beginning Data Science with R*,
DOI 10.1007/978-3-319-12066-9_1

- Statistical skills are essential in applying the right kind of statistical methodology along with interpreting the results.
- Programming skills are essential to implement the analysis methodology, combine data from multiple sources and especially, working with large-scale datasets.
- Domain expertise is essential in identifying the problems that need to be solved, forming hypotheses about the solutions, and most importantly understanding how the insights of the analysis should be applied.

Over the last few years, data science has emerged as a discipline in its own right.

However, there is no standardized set of tools that are used in the analysis. Data scientists use a variety of programming languages and tools in their work, sometimes even using a combination of heterogeneous tools to perform a single analysis. This increases the learning curve for the new data scientists. The R programming environment presents a great homogeneous set of tools for most data science tasks.

1.2 Why R?

The R programming environment is increasingly becoming a one-stop solution to data science. R was first created in 1993 and has evolved into a stable product. It is becoming the de facto standard for data analysis in academia and industry.

The first advantage of using R is that it is open source software. It has many advantages of other commercial statistical platforms such as MATLAB, SAS, and SPSS. Additionally, R works on most platforms: GNU/Linux, OS X, Windows.

R has its roots in the statistics community, being created by statisticians for statisticians. This is reflected in the design of the programming language: many of its core language elements are geared toward statistical analysis. The second advantage using R is that the amount of code that we need to write in R is very small compared to other programming languages. There are many high-level data types and functions available in R that hide the low-level implementation details from the programmer. Although there exist R systems used in production with significant complexity, for most data analysis tasks, we need to write only a few lines of code.

R can be used both as an interactive or a noninteractive environment. We can use R as an interactive console, where we can try out individual statements and observe the output directly. This is useful in exploring the data, where the output of the first statement can inform which step to take next. However, R can also be used to run a script containing a set of statements in a noninteractive environment.

The final benefit of using R is the set of R packages. The single most important reason for the growing popularity of R is its vast package library called the Comprehensive R Archive Network, or more commonly known as CRAN.[1] Most statistical analysis methods usually have an open source implementation in the form of an R package. R is supported by a vibrant community and a growing ecosystem of package developers.

[1] At the time of writing, CRAN featured 5763 packages.

1.3 Goal of This Book

Due to its statistical focus, however, R is one of the more difficult tools to master, especially for programmers without a background in statistics. As compared to other programming languages, there are relatively few resources to learn R. All R packages are supported with documentation; but it is usually structured as reference material. Most documentation assumes a good understanding of the fundamentals of statistics.

The goal of this book is to introduce the readers to some of the useful data science techniques and their implementation with the R programming language. In terms of the content, the book attempts to strike a balance between the *how*: specific processes and methodologies, while also talking about the *why*: going over the intuition behind how a particular technique works, so that the reader can apply it to the problem at hand.

The book does not assume familiarity with statistics. We will review the prerequisite concepts from statistics as they are needed. The book assumes that the reader is familiar with programming: proficient in at least one programming language. We provide an overview of the R programming language and the development environment in the Appendix.

This book is not intended to be a replacement for a statistics textbook. We will not go into deep theoretical details of the methods including the mathematical formulae. The focus of the book is practical; with the goal of covering how to implement these techniques in R. To gain a deeper understanding of the underlying methodologies, we refer the reader to textbooks on statistics [4].

The scope of this book is not encyclopedic: there are hundreds of data science methodologies that are used in practice. In this book we only cover some of the important ones that will help the reader get started with data science. All the methodologies that we cover in this book are also fairly detailed subjects by themselves: each worthy of a separate volume. We aim to cover the fundamentals and some of the most useful techniques with the goal of providing the user with a good understanding of the methodology and the steps to implement it in R. The best way to learn data analysis is by trying it out on a dataset and interpreting the results. In each chapter of this book, we apply a set of methodologies to a real-world dataset.

Data science is becoming ubiquitous: it is finding application in every domain area. In this book, we do not focus on any single domain such as econometrics, genetics, or web data. Our goal is to to provide tools to analyze any kind of data. When dealing with issues specific to a particular domain, we refer the reader to other books on data analysis available in the UseR! series.

1.4 Book Overview

In Chapter 2, we provide an overview of the R programming language. The readers who are already familiar with R can skim through it.

The remainder of the book is divided into individual chapters each covering a data science methodology. In Chap. 3, we start with getting the data into R and look at some of the data preprocessing tasks. In Chap. 4, we look at data visualization techniques. Next, in Chap. 5, we look at exploratory data analysis which can tell us about how the data is distributed. Chapters 6—Regression and 7—Classification, form the core of the book. In Chap. 6, we look at creating statistical models to predict numerical variables, while in Chap. 7, we look at analogous statistical models to predict categorical or class variables. In Chap. 8, we look at analyzing unstructured text data.

References

1. Big data, for better or worse: 90% of world's data generated over last two years. www.sciencedaily.com/releases/2013/05/130522085217.htm, May 2013.
2. Patil, D. J. (2012). *Data Jujitsu: The art of turning data into product*. O'Reilly Radar.
3. Drew Conway. The data science venn diagram. http://drewconway.com/zia/2013/3/26/the-data-science-venn-diagram.
4. Casella, G., & Berger, R. (2001). *Statistical inference*. Duxbury Resource Center.

Chapter 2
Overview of the R Programming Language

In this chapter we present a brief overview of the R with the goal of helping readers who are not very familiar with this language to get started.

We start with how to install R and use the development tools. We then look at the elements of the R programming language such as operators and data types. We also look at the syntax of different structures such as conditional statements and loops, along with functions. The R programming language has its share of peculiarities; we highlight some of them below. We look at installing and loading R packages from Comprehensive R Archive Network (CRAN). Finally, we will discuss the running R code outside the R console using Rscript.

2.1 Installing R

R is available on most computing platforms such as Windows, GNU/Linux, Mac OS X, and most other variants of UNIX. There are two ways of installing R: downloading an R binary and compiling R from source. The compiled binary executables for major platforms are available for download from the CRAN website http://cran.rstudio.com/.

R can also be installed using package managers. Ubuntu users can install R using the apt-get package manager.

```
$ sudo apt-get install r-base r-base-dev
```

Similarly, on Mac OS X we can install R using ports.

```
$ sudo port install R
```

Being an open source software, the source code for R is also available for download at http://cran.rstudio.com/. Advanced users can create their own binary executables by directly compiling from source as well.

There is an active community of R developers which regularly releases a new version of R. Each version also has a name assigned to it. Except for major releases, the versions of R are usually backward compatible in terms of their functionality. In this book, we use R version 3.1.0.

M. A. Pathak, *Beginning Data Science with R*,
DOI 10.1007/978-3-319-12066-9_2

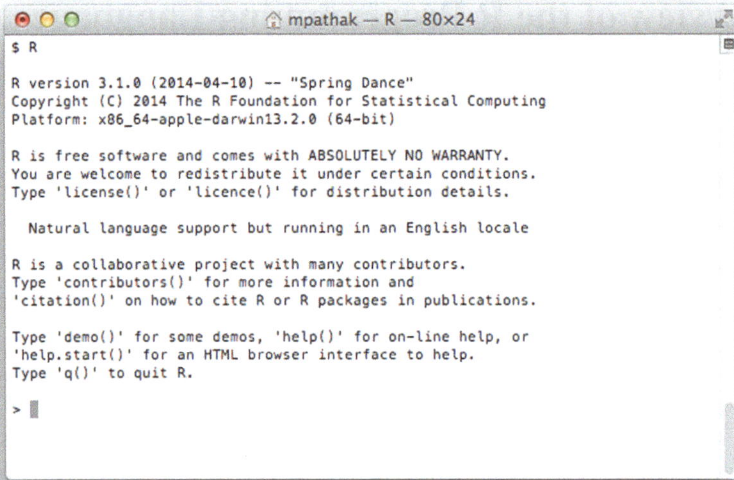

Fig. 2.1 R command-line application

2.1.1 Development Tools

The default installation of R on UNIX-based operating systems is in the form of a command-line application called `R`. We can start this application from a terminal to get a prompt as shown in Fig. 2.1. We can start typing R code at this prompt and see the output in the terminal itself. Alternatively, we can write the R code in an external text editor, and import or *source* the code in the command-line application. Many text editors such as Emacs and vi have R plugins that provide syntax highlighting and other tools. Although simplistic, this is the primary development environment for many R developers.

On Windows, R is installed as a graphical user interface (GUI) application (Fig. 2.2) with a set of development tools, such as a built-in editor. A similar GUI application called R.app also exists for Mac OS X. However, these applications are fairly basic when compared to integrated development environments (IDEs) for other programming languages.

Recently, there are many independent IDEs available for R. One of the most powerful IDEs is RStudio[1]. It is available in two editions: a GUI desktop application that runs R on the local machine, and server, where R runs on a remote server and we can interact with it via a web application.

[1] http://www.rstudio.com/.

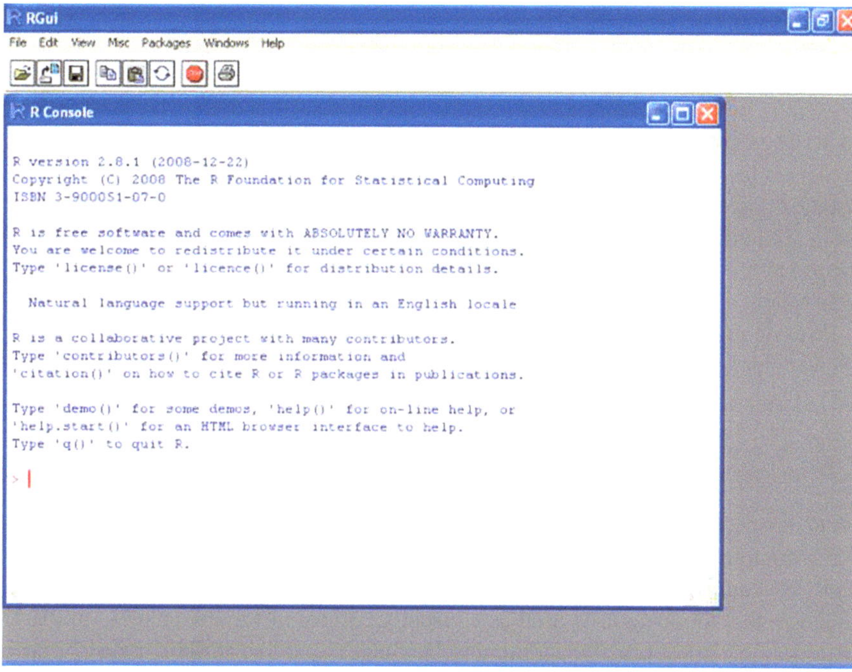

Fig. 2.2 R on Windows

2.2 R Programming Language

In this section, we briefly overview the R programming language. R is an interpreted language; the expression specified in an R program are executed line by line similar to other interpreted languages such as python or ruby rather than compiling the source code to an executable, as in C++. R is dynamically typed, which means that R infers the data types of the variables based on the context. We do not need to declare variables separately.

R has a bit of esoteric syntax as compared to most other programming languages. We look at the basic features of R below.

2.2.1 Operators

R provides arithmetic operators such as addition (+), subtraction (-), multiplication (*), division (/), and exponentiation (^). We enter the expression `3 + 4` at the R console. We do not need to terminate the expressions or lines by a semicolon.

```
> 3 + 4
[1] 7
```

As expected, R returns the answer as 7.

We use an assignment operator to assign the value of an expression to a variable. R has two assignment operators: the conventional assignment operator = which is present in most programming languages, and arrows <- and -> which are specific to R. The expression `x = 5` assigns the value 5 to x; the expression `x <- 5` and `5 -> x` have exactly the same effect. Historically, the arrow operator has been used for assignment in R. However, we do not have to use the arrow operator; apart from nonfamiliarity and typing twice as many characters, <- can also be a cause for errors: `x <- 5` vs. `x < -5`. Throughout this book, we only use the conventional assignment operator (=).

We can create expressions using variables. For instance, we assign the value 5 to the variable x and evaluate the square of x using the exponentiation (^) operator.

```
> x = 5
> x^2
[1] 25
```

R has peculiar syntax when it comes to variable names. The dot character (`.`) has a completely different meaning as compared to other programming languages. In R, we can use (`.`) in the variable names, so `x.1` and `x.2` are perfectly valid. In practice, the dot operator is used as a visual separator in variable names, similar to underscore in most other programming languages, e.g., the variable `prev_sum` is conventionally written as `prev.sum`. Underscores can also be used in R variable names as well, although they are less frequently used in practice as compared to dot.

2.2.2 *Printing Values*

Entering an expression on the R console evaluates the expression and then prints its value. Internally, R is calling the `print()` function with the value of the expression. We can call `print()` explicitly as well. This is useful when we want to print values for variables in a script.

```
> print(3 + 4)
[1] 7
```

Note that in the example in the previous subsection, the line `x = 5` did not print anything. R does not print the values of expressions when using the assignment operator.

The `print()` function prints the value of the expression and a new line. However, it is not straightforward to print multiple values using print(), for instance if we want to print the name of the variable along with the value. We need to print the output of the paste() function that concatenates two strings with a space.

```
> print(paste('the sum is', 3 + 4))
[1] "the sum is 7"
```

There is a simpler function called `cat()` that can print a list of values, so we do not require to call `paste()`. As the `cat()` function does not print the newline character, we need to specify it manually.

```
> cat('the sum is', 3 + 4, '\n')
the sum is 7
```

2.2.3 *Basic Data Types*

There are two kinds of data types: scalars that represent single-valued data or composite that represent collections of scalar data. Here we look at the scalar data types in R; we will discuss about the composite data types such as vectors and data frames in Chap. 3.

R provides multiple scalar data type formats such as numeric, integer, character, logical, and complex. The numeric data type is used to represent floating point numbers while integer data for representing only integer values. We can convert variables from numeric to integer using the `as.integer()` function.

```
> as.integer(2.56)
[1] 2
```

By default, R uses the numeric data type for integer values as well. We identify the data type of a variable using the `class()` function.

```
> x = 5
> class(x)
[1] "numeric"
```

We can also check if a variable is an integer using the is.integer() function. Such functions, as.datatype() and is.datatype() exist for all the data types mentioned above.

The character data type is used to represent strings. Unlike C or Java, R does not make a distinction between the single character char data type and multicharacter string data type. Additionally, we can use both single and double quotes to enclose strings; in this book, we primarily use single quotes.

```
> s1 = "string"
> s1
[1] "string"
> s2 = 'also a string'
> s2
[1] "also a string"
```

We convert between character and numeric variables using the `as.character()` and `as.numeric()` functions.

```
> as.character(2.5)
[1] "2.5"
```

```
> as.numeric('2.5')
[1] 2.5
```

Similar to other programming languages, R also has standard string processing functions such as computing the length of a string, finding substrings, splitting a string based on a character. The `stringr` library also provides a more consistent and easier to use set of functions for string processing.

The logical data type represents the boolean values: true and false. R uses two constants `TRUE` and `FALSE` to represent boolean values. These values are also represented by abbreviated constants `T` and `F`. In this book, we use these abbreviated constants to represent boolean values. R provides the standard boolean operators: and (`&`), or (`|`), not (`!`) along with relational operators such as equal to (==), less than (<) and greater than (>) that operate on numeric variables and return boolean values.

R also provides support for representing complex variables that contain a real and imaginary component.

```
> z = 2 + 3i
```

We can directly perform operations on the complex variables.

```
> z^2
[1] -5+12i
```

We will not be using complex data types in this book.

2.2.4 *Control Structures*

R provides control structures such as conditional branches (if-else) and loops. The syntax for if-else is similar to most other programming languages:

```
> if (x > 0) {
+      y = 'positive'
+ } else {
+      y = 'negative or zero'
+ }
```

In this case, `y` will be assigned the string `'positive'` if `x > 0` and `'negative or zero'` otherwise.

There are many other ways to write the same statement in R. Firstly, we can use if-else to return a value.

```
> y = if (x > 0) 'positive' else 'negative or zero'
```

We can also write the same expression using the `ifelse()` function, where the first argument is the boolean condition, and the second and third arguments are the respective values for the condition being true and false.

```
> y = ifelse(x > 0, 'positive', 'negative or zero')
```

An extension of the `ifelse()` function to multiple values is the `switch()` function.

R also provides multiple looping structures as well. The simplest loop is the while loop where we specify the boolean condition along with a body of steps that that are executed each time until the condition is met. The syntax for while loop is not different from that in C. We use the while loop to compute the sum of squares from 1 to 10.

```
> total = 0
> i = 1
> while (i <= 10) {
+       total = total + i^2
+       i = i + 1
+ }
> total
[1] 385
```

There are no ++ or += operators in R.

Another useful looping construct is the repeat loop, where there is no boolean condition. The loop continues until a break condition is met; conceptually, the repeat loop is similar to `while (T)`. We compute the same sum of squares from 1 to 10 using a repeat loop.

```
> total = 0
> i = 1
> repeat {
+       total = total + i^2
+       if (i == 10) break
+       i = i + 1
+ }
> total
[1] 385
```

R also has a powerful for loop that is more similar to for loop of python or Javascript as compared to the for loop in C. In this loop, we iterate over a vector of elements. We use the `in` operator to access an element of this vector at a time. We will discuss vectors in more detail in Chap. 3; for now, we construct a vector of elements from 1 to 10 as `1:10`. We compute the same sum of squares from 1 to 10 using a for loop below.

```
> total = 0
> for (i in 1:10) {
+       total = total + i^2
+ }
> total
[1] 385
```

2.2.5 Functions

R provides strong support for creating functions. In practice, most of the interactions we have with R is through functions: either provided by the base package, or user defined functions containing application logic.

The syntax for calling a function in R is similar to most other programming languages. For instance, we use the function `sum()` to compute the sum of a vector of numbers. This function is provided by the base package.

```
> sum(1:10)
[1] 55
```

R has special syntax for defining functions. As with other programming languages, we specify the function name, parameters along with body of the statements containing a return value. The difference is that we create a function using the `function` keyword and assign it to a variable. We will later see the reason for this.

We create a function called `avg()` to compute the average value of a vector of numbers. This is an implementation of the `mean()` function of the base package. We use the `length()` function that computes the number of elements or length of a vector.

```
> avg = function(x) {
+      return(sum(x)/length(x))
+ }
> avg(1:10)
[1] 5.5
```

We do not need to specify the return value explicitly in function; the last evaluated expression is automatically considered as the return value. For one line functions, we do not need to enclose the function body in braces. We can rewrite the same function as:

```
> avg = function(x) sum(x)/length(x)
> avg(1:10)
[1] 5.5
```

In R, functions are treated as first-class objects similar to other data types like numeric, character, or vector. This is a fundamental property of *functional programming* languages. Although the functional programming paradigm always had a strong niche community, it has become mainstream with the recent widespread adoption of languages like Scala, Clojure, OCaml,

The literal name of the function, in our case `avg`, corresponds to the function object, while the function call `avg(1:10)` corresponds to a value that is returned by the function when it is evaluated with the input `1:10`.

We can also assign the function sum to another variable `my.function`. Calling `my.function()` has the same effect of calling `sum()`.

```
> my.function = sum
```

```
> my.function(1:10)
[1] 55
```

Additionally, we can also pass a function as an argument to other higher-order functions. For instance, we create a function `sum.f()` that computes the sum of a vector after the function `f()` has been applied to it. Such a higher-order function should work for any function `f()`; at the time of defining `sum.f()`, we do not know what `f()` is going to be.

```
> sum.f = function(x, f) sum(f(x))
```

We compute the sum of squares of numbers between 1 and 10 using `sum.f()` below. As `f()`, we pass the square function.

```
> sum.f(1:10, function(x) x^2)
[1] 385
```

We created the function `function(x) x^2` without assigning any name to it prior to the function call. Such functions are known as anonymous functions.

R provides strong support for functional programming. There are benefits to using this paradigm including concise and cleaner code.

2.3 Packages

A self-contained collection of R functions is called as a package. This is especially useful when providing this package to other users. An R package can also contain datasets along with the dependencies. It is straightforward to create packages for our own R functions [1]. In this section we look at installing and loading other packages.

When we start R, the base package is already loaded by default. This package contains basic functions for arithmetic, input/output operations, and other simple tasks. The real power of R lies in its package library. There are thousands of packages available in the CRAN covering a very large number of data analysis methods. We can install these packages from R using the `install.packages()` function. This function downloads the source files for a package from a CRAN mirror website, builds the package, and stores the package in a local repository. The user does not need to do much for installing a package except for choosing the mirror.

As an example, we install the `stringr` package below. The `install.packages()` function provides a list of mirrors located around the world. We choose the first option: 0-Cloud.

```
> install.packages('stringr')
Installing package into /Users/mpathak/Library/R/3.1/library
(as lib is unspecified)
--- Please select a CRAN mirror for use in this session ---
CRAN mirror

 1: 0-Cloud                          2: Argentina (La Plata)
```

```
 3: Argentina (Mendoza)           4: Australia (Canberra)
 5: Australia (Melbourne)         6: Austria
 7: Belgium                       8: Brazil (BA)
 9: Brazil (PR)                  10: Brazil (RJ)
11: Brazil (SP 1)                12: Brazil (SP 2)
13: Canada (BC)                  14: Canada (NS)
15: Canada (ON)                  16: Canada (QC 1)
17: Canada (QC 2)                18: Chile
19: China (Beijing 1)            20: China (Beijing 2)
21: China (Hefei)                22: China (Xiamen)
23: Colombia (Bogota)            24: Colombia (Cali)
25: Czech Republic               26: Denmark
27: Ecuador                      28: France (Lyon 1)
29: France (Lyon 2)              30: France (Montpellier)
31: France (Paris 1)             32: France (Paris 2)
33: France (Strasbourg)          34: Germany (Berlin)
35: Germany (Bonn)               36: Germany (Goettingen)
37: Greece                       38: Hungary
39: India                        40: Indonesia (Jakarta)
41: Indonesia (Jember)           42: Iran
43: Ireland                      44: Italy (Milano)
45: Italy (Padua)                46: Italy (Palermo)
47: Japan (Hyogo)                48: Japan (Tokyo)
49: Japan (Tsukuba)              50: Korea (Seoul 1)
51: Korea (Seoul 2)              52: Lebanon
53: Mexico (Mexico City)         54: Mexico (Texcoco)
55: Netherlands (Amsterdam)      56: Netherlands (Utrecht)
57: New Zealand                  58: Norway
59: Philippines                  60: Poland
61: Portugal                     62: Russia
63: Singapore                    64: Slovakia
65: South Africa (Cape Town)     66: South Africa(Johannesburg)
67: Spain (A Corua)            68: Spain (Madrid)
69: Sweden                       70: Switzerland
71: Taiwan (Taichung)            72: Taiwan (Taipei)
73: Thailand                     74: Turkey
75: UK (Bristol)                 76: UK (Cambridge)
77: UK (London)                  78: UK (London)
79: UK (St Andrews)              80: USA (CA 1)
81: USA (CA 2)                   82: USA (IA)
83: USA (IN)                     84: USA (KS)
85: USA (MD)                     86: USA (MI)
87: USA (MO)                     88: USA (OH)
89: USA (OR)                     90: USA (PA 1)
91: USA (PA 2)                   92: USA (TN)
93: USA (TX 1)                   94: USA (WA 1)
95: USA (WA 2)                   96: Venezuela
97: Vietnam

Selection: 1
trying URL 'http://cran.rstudio.com/src/contrib/
stringr_0.6.2.tar.gz' Content type 'application/x-gzip'
length 20636 bytes (20 Kb) opened URL
```

```
==================================================
downloaded 20 Kb

* installing *source* package stringr ...
** package stringr successfully unpacked and MD5 sums checked
** R
** inst
** preparing package for lazy loading
** help
*** installing help indices
** building package indices
** testing if installed package can be loaded
* DONE (stringr)
```

The messages from `install.packages()` function confirm that `stringr` package was installed correctly.

We need to call `install.packages()` for a function only once. After starting R, we use the `library()` function to load the packages that we need into our workspace. We load the `stringr` package below.

```
> library(stringr)
```

An alternative to `library()` is the `require()` function which has the same syntax.

We use various packages throughout this book. For each package, we only use the `library()` function to load it assuming that it has been installed. If that package has not installed already, we require the reader to install it using the `install.packages()` function.

2.3.1 R Help System

Another advantage of using R is that we can access documentation directly from the R console. For any function, we can look up the documentation using the `help()` function. We look up the documentation for the `cat()` function below.

```
> help(cat)
```

Alternatively, we can also look up the documentation using the `?` operator.

```
> ?cat
```

Along with documentation, we can also see the example usage of a function using the `example()` function.

```
> example(cat)

cat> iter <- stats::rpois(1, lambda = 10)

cat> ## print an informative message
```

```
cat> cat("iteration = ", iter <- iter + 1, "\n")
iteration =  8

cat> ## 'fill' and label lines:
cat> cat(paste(letters, 100* 1:26), fill = TRUE,
         labels = paste0("{", 1:10, "}:"))
{1}: a 100 b 200 c 300 d 400 e 500
{2}: f 600 g 700 h 800 i 900 j 1000
{3}: k 1100 l 1200 m 1300 n 1400 o 1500
{4}: p 1600 q 1700 r 1800 s 1900 t 2000
{5}: u 2100 v 2200 w 2300 x 2400 y 2500
{6}: z 2600
```

2.4 Running R Code

In the above examples, we have exclusively used R in the interactive mode, mostly by typing in commands into the R console. This is useful for quickly trying out single R functions and exploring the data. This becomes cumbersome when we have multi-line block of statements or functions that we want to reuse. In this case, we write the R code in a text file, conventionally with the `.r` extension. This has the added benefit of storing our R code for future use.

As an example, we create the file `test.r` with the code to sum the squares of numbers from 1 to 10.

```
# loop.r
total = 0
for (i in 1:10) {
    total = total + i^2
}

print(total)
```

We use the `source()` function to run the code from `loop.r` in the R console.

```
> source('loop.r')
[1] 385
```

We need to explicitly use `print()` to print values when using `source()`.

We can also run the R code without starting the R console using the tool Rscript. We pass the filename containing the source code as the command-line argument to Rscript.

```
$ Rscript loop.r
[1] 385
```

We can also pass additional command line arguments to Rscript which we can access inside `loop.r`.

Reference

1. Leisch, F. (2009). Creating R packages: A tutorial.
http://cran.r-project.org/doc/contrib/Leisch-CreatingPackages.pdf.

Chapter 3
Getting Data into R

In this chapter, we look at reading the data into R, which is usually the first data analysis task. It is convenient to read data into R due to its support for reading data from a wide variety of formats.

The basic data types in R are designed to make data analysis convenient. Similar to most programming languages, R has support for numeric, character, string, boolean data types. What makes R attractive are the structured data types; in this chapter, we look at two of these: vectors and data frames. Vectors are simply a sequence of elements that are of basic data types such as numeric or character. Data frames are an abstraction for tabular data: The rows represent the data points or observations and the columns represent the variables or the aspects of the data that we are measuring. Data frame is the fundamental data type in R as it is used by most of the standard functions and even across other packages. In this book, we use data frames extensively to store the data. There are many useful functions to slice and dice the data frames which we will discuss in this chapter.

The simplest format is delimiter-separated files such as comma separated values (CSV). In enterprise environments, the data is usually stored in databases. There are R packages that make it easy to read the data from databases as well. In this chapter, we will look at reading the data from a PostgreSQL database into R using SQL.

The quality of the data is a critical factor in any kind of data analysis. We cannot derive any reasonable inference from the data if the data itself is messy or erroneous. Often enough, the messiness of the data is rather subtle; this introduces biases in our analysis which is hard to detect, and cause problems later. Before we get started with performing the analysis, it is important to look at the data for such errors.

In this chapter we will also look at the three types of variables: numeric, categorical, and ordinal. First, a common source of error is misidentifying the data type for a variable, where we assume a categorical variable such as zip code to be numeric. Second, most nontrivial datasets contain some amount of data entry errors. We will discuss how to identify and correct some of the common patterns of these errors. Finally, most real world datasets also contain missing values. Removing the parts of the data with missing values reduces the size of the dataset. A more effective strategy

M. A. Pathak, *Beginning Data Science with R*,
DOI 10.1007/978-3-319-12066-9_3

Fig. 3.1 Comma-separated values (CSV) file: cars.csv

```
make,model,trim,year,price,mpg
Honda,Civic,LX,2012,17633,28
Honda,Civic,LX,2013,19346,28
Honda,Accord,EX-L,2013,25609,21
Toyota,Corolla,LE,2013,16821,26
Aston Martin,DB9,Coupe,2012,189915,12
```

is data imputation, where we replace the missing values based on the context. At the end of the chapter, we will discuss simple data imputation strategies.

3.1 Reading Data

The first step of any data analysis is to read data into R. In many cases, the data in its original form is not be directly usable for analysis. This is typical in cases where the data is in an unstructured format such as nontabular plain text or a semi-structured format such as web pages. Sometimes, the data might be in nontextual format such as scanned documents.

In such cases, it is necessary to transform the data into a more usable representation. This task broadly falls under the category of information extraction. There are two options to convert the data in a nontextual format to a machine readable format: automated optical character recognition (OCR) or using manual transcription. It is becoming common to use crowdsourcing tools such as Amazon Mechanical Turk and Servio to transcribe the data with high accuracy and low cost. For machine-readable documents, automated information extraction algorithms deal with converting information from unstructured sources to a structured form.

Once the data is in a tabular structure, it is convenient to read it into R given its support for a variety of data formats.

3.1.1 Text Files

The simplest format to read data into R is using delimited text files such as CSV and tab-separated values (TSV). These files are used by most spreadsheet applications, databases, and statistical analysis tools either as a base file format or as a data export file format.

A delimiter-separated values file, e.g., Fig. 3.1, contains tabular data: both numbers and text, with each row per line separated by the newline character, and each column-entry of that row separated by the delimiter character. In our example, the columns (`make`, `model`, `trim`, `year`, `price`, `mpg`) are the variables that we are recording about the data. The rows of the file are the observations: individual entries for the variables. Conversely, we can think of a variable as a vector of observations, and the data as a list of variables.This is similar to how data is represented in a relational database table.

In a delimiter-separated file, the first row of the file is usually a header row containing the names of the columns. We use the `read.table()` function to read such files; we use it to read a tab-separated file `cars.txt` below.

```
> data = read.table('cars.txt',sep=',')
```

We need to provide a delimiter value to the `read.table()` function. A similar function is `read.csv()` that assumes the delimiter value is ',' which makes it convenient to read CSV files. This function uses the same parameters as `read.table()` function.

The `read.table()` function by default assumes that the first row of the file contains the header. If the file does not contain the header, this function would assume the first data row to contain the column names which leads to undesirable effects. Instead of having a column named `make`, we will end up with a column named `Honda`. To ensure R does not use the first row as the header, we call `read.table()` with `header = F` as a parameter.

```
> data = read.table('cars.txt',header=F,sep=',')
```

3.1.1.1 Data Frames

The `read.table()` function returns an object of the type `data.frame`. The `data.frame` data type captures the tabular representation of the data: it contains individual observations or entries as rows and variables as columns. We can also view a data frame as a list of variables with individual observations as rows. Data frame a fundamental data type in R; most of the standard functions use data frames as input parameters and return data frames as well.

It is convenient to access the data from the `data.frame`. In the data variable that we read above from cars.csv, we can access the first row by indexing on the first dimension.

```
> data[1,]
   make model trim year price mpg
1 Honda Civic   LX 2012 17633  28
```

Vector is another fundamental data type in R. A vector is just a sequence of numbers; the columns of a data frame are vectors as well. There are multiple ways to construct a vector in R: the `c()` function creates a vector containing the arguments it is called with, e.g., `c(1,1,2,3,5)` gives us a vector containing 1, 1, 2, 3, and 5. This function works on strings too. If we need a vector of consecutive numbers, we can use the column operator; 1:10 is shorthand for `c(1,2, ... ,10)`. The `seq(i,j,step)` function returns a vector of numbers from `i` to `j` that is separated by `step`, e.g., `sep(1,10,2)` creates a vector containing odd numbers between 1 and 10. The `rep(x,n)` function creates a vector containing x repeated n times.

Vector is a useful data type; we can use vectors to index data frames, e.g., we can access the first three rows using `data[1:3,]` or the first, third, and fifth row by `data[c(1,3,5),]`.

To access a column, we can index on the second dimension, e.g., `data[,1]` for the first column. However, a more intuitive way to access the columns is by using the `$` operator: `data$column_name` returns the appropriate column on the data frame. For example, to access the column `mpg` using the `$` operator, we can use `data$mpg`.

```
> data$mpg
[1] 28 28 21 26 12
```

If you do not prefer the `$` operator, there are alternative ways to access a data frame column by name. We list some of them below.

```
> data[['mpg']]
[1] 28 28 21 26 12
> data[,'mpg']
[1] 28 28 21 26 12
```

When working with data frames, another useful function is `attach()`. This function adds the column names to the R search path, so we can access them directly, without using the `$` operator. This is a convenient shorthand when we are using the same column names over and over again.

```
> attach(data)
> mpg
[1] 28 28 21 26 12
```

Readers familiar with Matlab, Octave, or the numpy package in Python would immediately compare data frames with matrices. In fact, R does have a separate matrix data type; there are certain similarities between matrices and data frames. An important difference between the two data types is that in a data frame, the columns can contain data of different data types. You can have one column with strings: model in our example, and one with numbers: mpg. It is easy to switch between matrix and data frame types by using the `as.matrix()` and `as.data.frame()` functions.

As compared to matrices, data frames are more convenient to use, given that we can index them by column name and, also apply different formatting to individual columns. In this book, we will primarily be using data frames to represent our data.

We can also use the `read.table()` function to read a remote delimiter-separated text file by calling it with a URL. This makes it easy to access a dataset that is published on the web, without the additional step of downloading the file locally.

R also has a set of functions in the `foreign` package that can access data files exported by other statistics software such as Octave, Stata, SPSS, SAS (see Table 3.1.

R also provides analogous functions `write.csv()` and `write.table()`, that write data frames to files. The default arguments to this function outputs the row numbers as well; to disable that, we need to specify the argument row.names=F, e.g.,

```
> write.csv(data,'output.csv',row.names=F)
```

Table 3.1 R functions to read data from other statistics software

read.dbf	Read a DBF File
read.dta	Read Stata binary files
read.epiinfo	Read Epi Info data files
read.mtp	Read a Minitab Portable Worksheet
read.octave	Read Octave Text Data Files
read.spss	Read an SPSS data file
read.ssd	Read a data frame from a SAS Permanent Dataset
read.systat	Read a data frame from a Systat File
read.xport	Read a SAS XPORT Format Library

3.1.1.2 Reading Data from Databases

In most large scale data analysis projects, the data is stored in some kind of a database.[1] This is usually the case when working in an enterprise. This is because databases are built for managing large scale data well; databases provide a principled way to store the data and corresponding mechanisms to retrieve it.

Although the dominance of SQL-based relational databases has been challenged in recent years by NoSQL solutions, they still remain widely used for storing tabular data. The availability of high quality open source database solutions such as PostgreSQL and MySQL has been helpful for this.

R provides extensive support for accessing data stored in various SQL databases. On one side, there are packages such as `DBI` and `RODBC` for connecting to generic database drivers, while on the other side, there are packages for tailored to individual databases such as `RPostgreSQL`, `RMySQL`, `ROracle`, and `RSQLite`. The API for most of these packages is not very different. R does have good support for accessing data from NoSQL databases such as MongoDB using `RMongo` package.

We briefly look at accessing data from a SQL database below. As an example, let us assume that we have the data from our file cars.txt stored in a table called `cars` in PostgreSQL. In the PostgreSQL console, we can inspect this table.

```
testdb=# select * from cars;
     make      |  model  | trim  | year | price  | mpg
---------------+---------+-------+------+--------+-----
 Honda         | Civic   | LX    | 2012 |  17633 |  28
 Honda         | Civic   | LX    | 2013 |  19346 |  28
 Honda         | Accord  | EX-L  | 2013 |  25609 |  21
 Toyota        | Corolla | LE    | 2013 |  16821 |  26
 Aston Martin  | DB9     | Coupe | 2012 | 189915 |  12
(5 rows)
```

[1] Readers not familiar with databases and SQL can skip this section without loss of continuity.

We access this data in the code snippet below. We first obtain a driver from `dbDriver()` function and then use it to open a connection to the database using the `dbConnect()` function. We need to provide the database access credentials here. We then use the `dbSendQuery()` function to send the SQL query `select * from cars where make = 'Honda'` to the connection. The `dbSendQuery()` function returns the output of the query into a result set object. We then obtain a data frame corresponding to the result set using the `fetch()` function. Passing `n = -1` returns all the rows in the result set; passing `n = 2` will return only the last two elements.

```
library(RPostgreSQL)

# create an RPostgreSQL instance
drv = dbDriver('PostgreSQL')

# open the connection
conn = dbConnect(drv, user = 'username', password =
   'password', dbname = 'testdb', host = 'localhost')

# Run an SQL statement by creating first a result set
object rs <- dbSendQuery(conn,
       statement = 'select * from cars where make =
       \'Honda\'')

# we now fetch all rows from the result set into a
  data frame db.data <- fetch(rs, n = -1)

dbDisconnect(conn)
```

As expected, the data frame contains the output of the SQL query, with the columns being the variables and rows being the observations.

```
> print(db.data)
   make  model trim year price mpg
1 Honda  Civic LX 2012 17633   28
2 Honda  Civic LX 2013 19346   28
3 Honda Accord EX-L 2013 25609 21
```

RPostgreSQL has many other useful functions. `dbWriteTable (conn,'new_table',data)` writes a data frame to a table and returns a true value if the write is successful.

In the above example, we performed an operation in the SQL query by using the where clause to find only the rows with `make = 'Honda'`. We could have obtained the entire data frame and performed the same computation in R. Whether to perform this computation in SQL or R is a question of efficiency. For large tables, it is usually more efficient to do simple computation such as filtering rows and columns, or aggregation on the database side using SQL. If do the same computation in R, we would need to retrieve the entire table, send it over the network, store it into memory in R, and then iterate over it. Having an index on the column would make the filtering operation orders of magnitude faster when performed in the database.

Using stored procedures takes this view to an extreme, here the business logic, in our case, the statistical analysis itself can be performed inside the database. This provides high processing and bandwidth efficiency as it eliminates the need most of the intermediate data transfers. PostgreSQL has support for PL/R, which is a language for writing stored procedures using R [1].

The `sqldf` package in R takes the diametrically opposite view. This package provides a mechanism to query R data frames using SQL. It does so by first loading the data frame into a temporary database table, executing the query against it, and deleting the temporary table returning the results. `Sqldf` can connect to many SQL databases like PostgreSQL; in the absence of a database, it uses SQLite.

`Sqldf` can also query text files without first loading them into data frames. We can query the data frame containing the cars data to find the average mpg permake type.

```
> library(sqldf)
> sqldf('select make,avg(mpg)from data group by make')
          make avg(mpg)
1 Aston Martin 12.00000
2        Honda 25.66667
3       Toyota 26.00000
```

The sqldf function also returns another data frame containing the output of the query.

Interestingly, `sqldf` performs faster than the corresponding R operations in benchmark tests [3]. The utility of sqldf is mainly in its alternative syntax. Readers more comfortable with SQL find it productive to use as compared to the standard data frame functions.

3.2 Cleaning Up Data

After the data is loaded up into R, the next step is to look at data for errors. In the real world, data is usually messy; we cannot expect our data analysis to yield clear results if we use it directly. In this section we will look at how to identify and clean up the errors in the data arising due to data entry errors and missing values. As an example, we use a fictional dataset containing responses of a group of individuals to questions about their health and physical characteristics. This example dataset is a modification to the *survey* dataset in the MASS package.

Case Study: Health Survey

The dataset contains responses from 257 individuals to a basic health survey containing six questions: The individuals were asked about their sex, height, weight and whether they are left or right handed, whether they exercise and smoke.

This is included in the file `survey.csv` and the first few entries are shown in Fig. 3.2.

```
"sex","height","weight","handedness","exercise","smoke"
"Female",68,158,"Right","Some","Never"
"Male",70,256,"Left","None","Regul"
"Male",NA,204,"Right","None","Occas"
"Male",63,187,"Right","None","Never"
"Male",65,168,"Right","Some","Never"
"Female",68,172,"Right","Some","Never"
"Male",72,160,"Right","Freq","Never"
"Female",62,116,"Right","Freq","Never"
"Male",69,262,"Right","Some","Never"
"Male",66,189,"Right","Some","Never"
```

Fig. 3.2 survey.csv

3.2.1 Identifying Data Types

Let us begin by first looking at the variables: height and weight are numeric values or more generally are called real-valued or *quantitative* variables. These variables represent a measurable quantity. Quantitative variables occur everywhere in data analysis. Commonly found examples include per capita income, number of touchdowns a player scores in an year, or hat size for a set of individuals.

The sex and handedness variables take two values: male and female, and left- and right-handed, respectively. These are called *categorical* variables, *nominal* variables, or simply *factors*. These type of variables partition the data into disjoint categories, each represented by one of a fixed set of values; in the case of sex, the two categories would be the data points with sex male and those with sex female. Other examples of categorical variables include area or zip code where a person lives, race, blood group, or even words in a document. The main difference between numeric and categorical variables is that we can compare two values of a numeric variable: height 66" is more than 65", but we cannot similarly compare left- and right-handedness by themselves.

R represents categorical variables using the *factor* data type. A vector can be converted into a factor by using the `as.factor()` function. Factors in R are stored differently as compared to numeric variables. Internally, they are stored as a vector of integer values corresponding to a set of levels representing each category. This encoding enforces the factor variables to be treated differently in the data analysis. If the zip code variable is a factor, R would return an error if we tried to compute the average zip code.

The exercise and smoke variables are more subtle; they take values such as "None," "Some," "Frequent" for exercise, and "Never," "Occasionally," and "Regular." Although these variables also partition the data, there is an implied relationship between the groups: exercising frequently is more than exercising sometimes which is more than doing no exercise. These variables are called *ordinal* variables. Movie ratings (on a 0–5 scale), income groups are all examples of ordinal variables. The added structure of ordinal variables plays a key role for many statistical analyses.

```
> data$smoke
  [1] Never Regul Occas Never Never Never
   ...
Levels: Heavy Never Occas Regul
```

In R, a factor can be converted into an ordinal variable using the `ordered()` function. This function does not know the right order to apply, so it picks them alphabetically. To enforce the order that we have in mind, we can pass the levels vector as follows.

```
> data$smoke = ordered(data$smoke,
           levels=c('Never','Occas','Regul','Heavy'))
  [1] Never Regul Occas Never Never Never
  ...
Levels: Never < Occas < Regul < Heavy
```

3.2.2 *Data Entry Errors*

Our health survey dataset, like many other datasets, has been collected by a manual process. The data collection involves human participation at many points in the data collection process, and therefore introduces the possibility of data entry errors. Survey data is usually collected from physical media such as paper forms that are converted into digital format using OCR technology, phone or personal interviews, or from a software user interface. In most of such data collection processes, data entry errors do creep in.

The frequency of data entry errors depends on the nature of the task. In typing-based data entry tasks, the typical error rates are 1% per keystroke [4]. At the price of lower entry speeds, such errors can be reduced by using double entry checks: where two data entry operations work on the same task and compare their errors at the end.

Let us look at the sex variable in our dataset. Using our domain knowledge, we assume that this variable can only take "Male" or "Female" as values, and that it should be a factor. The values of sex variable as they appear in our dataset are somewhat more suspect: we use the `unique()` command to get a listof unique values.

```
> unique(survey$sex)
[1] Female Male  <NA>  F     M    male   female
    Levels: F female Female M male Male
```

The output of the `unique()` function tells us that the sex variable contains "Male" and "Female"; but also contains a few occurrences of their variants such as using lower case "female" or a single letter "F" instead of "Female" in the proper case. Such errors occur when different individuals who employ their own notation supervise different parts of the survey, or when the users do not fully comply with the data input guidelines. In other cases, such errors are compounded by the typos occurring during data entry. We see that `data$sex` has an entry with the value `NA` as well. We will discuss this in the next subsection.

In our data frame, however, the sex variable is a factor with six levels: F, female, Female, M, male, Male. Any analysis involving this variable will split the data into six categories, which is clearly an error we would want to avoid. We fix this problem by replacing the erroneous values with the correct ones. The `which()` function selects a subset of the entries of the variable matching a condition. For example, we find the entries where the variable `sex` has the value `'F'` by:

```
> which(data$sex == 'F')
[1] 210 211 212
```

The `which()` function indicates that these three rows have values `sex = 'F'`. To replace the values of these entries, we use the output of the which function as an index.

```
> data$sex[which(data$sex == 'F')] = 'Female'
```

We can also use the which function to slice the data over multiple variables using the boolean and & and boolean or | operators. For example, we obtain the data points with `sex = 'Male'` and `weight > 150` below.

```
> which(data$sex == 'Male' & data$weight > 150)
```

It is not necessary to use `which()` to index a data frame. We can use the boolean expression itself. In our case, the expression `data$sex == 'F'` returns a vector of true and false values of the same length as the number of rows in the data frame. When indexed with a data frame or vector, it returns the rows corresponding to the true value. We can replace the `sex = 'F'` values in the following way as well.

```
> data$sex[data$sex == 'F'] = 'Female'
```

3.2.3 *Missing Values*

In R, the missing values for a variable are denoted by the NA symbol. Both integer and character variables can take this value. The NA symbol is different from the NaN symbol, which indicates that the value is not a number as a result of an invalid mathematical operation, e.g., divide by zero. NaN indicates that the value is not a number but is present, whereas NA indicates that the value is absent from our data.

There are usually some missing values in any nontrivial data collection exercise. There are many reasons that could give rise to missing values. In some cases, users are not willing to provide all of their inputs. Also, variables that are personally identifiable information (PII), e.g., SSN, could be removed from the dataset for privacy purposes. This is common when the nature of the dataset is personal, such as in health care. In other cases, the inputs that are not recorded and stored properly give rise to missing data. If the data collection is automatic, missing data is often caused due to a sensor failure or data transmission and transcription errors.

The survey dataset contains a few missing entries for the variables sex, height, handedness, and smoke. R provides many useful functions to deal with missing data.

We cannot check if a variable contains NA values by comparing it with NA. This will only return a vector of NA's.

```
> data$height == NA
  [1] NA NA NA NA NA ...
```

`is.na()` is the appropriate function to use here. We find the entries containing missing values for the variable height by:

```
> which(is.na(data$height))
 [1]   3  12  15  25  26  29  31  35  58  68  70  81
   83   84  90  92  96 108 121
[20] 133 157 173 179 203 213 217 225 226
```

The standard R functions also have support for `NA`'s. As the height variable contains `NA`'s, computing `mean(data$height)` outputs `NA`. We obtain the mean of the nonmissing or observed values by using the `na.rm=T` flag.

```
> mean(data$height, na.rm = T)
[1] 67.89474
```

We can use the `na.fail()` function to check if a data frame or a contains missing values. If it contains any, the function returns an error, otherwise its output is the same as the input. Sometimes, we only want to consider the observed values, and ignore the missing values all together. The `na.omit()` and `na.exclude()` functions do just that.[2] We get a subset of a data frame or a vector that contains observedvalues by

```
data.observed = na.omit(data)
```

Removing the entries with missing values is also called list-wise deletion.

The above functions deal with performing statistical analysis while removing the missing values. As we are using only a subset of the data, this leads to a data loss. An alternative strategy is called imputation where we replace the missing data values by substituted values obtained using on other information.

Simple data imputation strategies include replacing the missing data for a numerical variable with a constant value. We usually choose the replacement value based on nature of the data. A missing value is equivalent to a zero in some cases, e.g., if we are asking for number of apple pies consumed on that day. The default value could be nonzero for some other cases. We replace the missing values by a constant value by

```
data$height[is.na(data$height)] = 60
```

For other variables such as height, replacing by an arbitrary constant value like 5 ft. might not be accurate. One reason is that it is not consistent with the data distribution. Statistical analysis of a large sample of heights for either gender shows

[2] There are only minor differences in na.omit() and na.exclude() that are related to the na.action parameter of the output that is relevant when performing regression.

that it follows a Gaussian distribution centered around the mean height [2]. It is more accurate to replace the missing height for males and females by their respective mean heights. We use the `which()` function to first find the mean female height, and then use it with the `is.na()` function to replace the missing values. We use the same strategy for males.

```
> female.height = mean(data$height[which(data$sex
    == 'Female')], na.rm=T)
> data$height[which(data$sex == 'Female'
    & is.na(data$height))] = female.height
```

For non-numerical categorical variables, there is no equivalent for the mean value. One strategy is to replace the missing value by the mode of the data, which is the most frequently occurring value which is also called the mode.

3.3 Chapter Summary

In this chapter, we first saw various ways to get the data into R. This usually involves conditioning the data so that it is a readily consumable tabular format. R has extensive support to load the data from delimiter-separated text files such as CSV and relational databases. Most of the R functions to load data return objects of type data frames. It is easy to access a subset of rows and columns of the data frame and select individual columns with the $ operator.

Once we have data frames loaded into R, the next step is to clean the data. In most nontrivial data analyses, the data suffers from various quality issues such as misidentified data types, data entry errors and missing values. It is important to identify and resolve these issues to maintain the quality of the analysis.

We discussed simple strategies for dealing with missing data such as replacing the missing values with data means and modes. There are many other data imputation strategies that use statistical models to infer the missing data values. Packages such as Amelia II, mitools, and Mice provide implementations for these strategies.

References

1. Conway, J. E. Pl/r - r procedural language for postgresql. http://www.joeconway.com/plr/. Accessed 1 Aug 2014.
2. Cook, J. D. (2008). Distribution of adult heights. http://www.johndcook.com/blog/2008/11/25/distribution-of-adult-heights/. Accessed 1 Aug 2014.
3. sqldf. http://code.google.com/p/sqldf/. Accessed 1 Aug 2014.
4. Table of Basic Error Rates in Many Applications. http://panko.shidler.hawaii.edu/HumanErr/Basic.htm. Accessed 1 Aug 2014.

Chapter 4
Data Visualization

4.1 Introduction

An important step in the data science methodology is obtaining a visual representation of the data. This has multiple advantages: first, as humans, we are better at extracting information from visual cues, so a visual representation is usually more intuitive than a textual representation. Second, data visualization, for the most part, also involves a data summarization step. A visualization provides a concise snapshot of the data. As it is often said, “a picture is worth a thousand words.”

The goal of data visualization is to convey a *story* to the viewer. This story could be in the form of general trends about the data or an insight. Creating effective data visualizations is similar to being good at storytelling: it involves conveying relevant and interesting information at the right level of detail. This makes the data visualization task more of an art and less of an exact science. Along with the core statistical insights that we intend to present, aesthetics of presentation are equally important. This is not to say that we can hide the lacunae of our analysis by clever presentation tricks. Just like clear writing, clear and effective presentation should make our data analysis transparent, so that the viewer can fully understand and appreciate the complete picture.

One of the major benefits of using R for data analysis is its extensive support for creating high-quality visualizations. The data visualizations created with R have appeared in the New York Times [1] and the Economist, and is standard fare for articles in scientific journals. There are a large number of R functions for visualizing almost all data types, from numerical to categorical data in multiple dimensions. There is also support for displaying the information in a wide variety of formats, from basic point and line plots to bar and pie charts displaying aggregated data, to more complex plots. We can easily control all aspects of the visualization programmatically, as the plotting functions are completely configurable and can be automated via scripting. This is often not the case with many other data visualization tools that use a point-and-click interface.

In this chapter, we look at different types of data visualizations that we can create using R. We start with the plotting functionality in the base R package, and then move to the more sophisticated, yet simple to use, ggplot2 package. We will look

M. A. Pathak, *Beginning Data Science with R*,
DOI 10.1007/978-3-319-12066-9_4

Table 4.1 List of variables in the teams data

Variable name	Description
team	Name of the MLB Team
code	Three-letter code name of the team
league	League the team plays in
division	Division the team plays in
games	Number of games played
wins	Number of wins
losses	Number of losses
pct	Winning percentage
payroll	Estimated total team payroll

at specialized plots for exploratory data analysis such as histograms in subsequent chapters. As a case study, we use a dataset from Major League Baseball (MLB).

Case Study: 2012 Major League Baseball Season For our case study, we look at the team statistics for the 2012 regular season of MLB.[1] The dataset contains the variables as listed in Table 4.1. This includes the team details, which league and division it plays in, the performance for the team in terms of number of wins and losses, and the estimated total payroll, which is the amount of money (in USD) that the team spent on its players.

In 2012, there were 30 teams in the MLB, each participating in either the American League (AL) or National League (NL). In each of these leagues, there are three divisions: East, West, and Central, making it a total of six divisions. In the regular season, each team played 162 games. The payroll of the teams is spread over a wide range, from a minimum of $ 38 million for Houston Astros to a maximum of $ 198 million for the New York Yankees.

In this case study, one natural question is whether teams with higher payrolls perform better. We aim to get some insight into that through our visualizations.

We load the dataset using read.csv(). We use the attach() function so that the individual variables are available as vectors.

```
> data = read.csv('teams.csv')
> attach(data)
```

4.2 Basic Visualizations

In this section, we look at creating visualizations using the default graphics package in R.

[1] Dataset obtained from http://www.baseball-reference.com.

4.2.1 *Scatterplots*

The plot() function in R is a versatile function; it can take vectors, a function, or any R object that has a plot() method as input. In its most simplest form, it takes two vectors as input to create a scatterplot. A scatterplot visualizes the relationship between two variables by simply displaying data points with x coordinates as one variable and y coordinates as another variable. Conventionally, the variable on the x axis is the independent variable and the variable on the y axis is thedependent variable.

Independent and Dependent Variables

In data analysis, there are independent variables and dependent variables. The independent variables are the inputs or something a team management can control, e.g., the team payroll. The dependent variables are the output, or something a team can observe but not directly control, e.g., the number of wins, except for a team that is intentionally losing games. There are also "other" variables that are neither independent nor dependent for the purposes of an analysis. One of the goals of data analysis is to identify the relationship between the dependent and independent variables.

We draw a scatterplot between payroll and wins in our dataset. Figure 4.1 shows the output.

```
> plot(payroll,wins)
```

The plot has payroll on the x axis and number of wins on the y axis. The data points in the plot are the 30 teams in our dataset. The most interesting part is the interpretation of the plot: there is no apparent linear relationship between the two variables. We have teams with lower payroll having more wins than teams with much higher payrolls. Intuitively, this indicates that simply increasing the payroll does not increase the number of wins. We can dig deeper into this; as we will see in later chapters, there are techniques such as correlation analysis and regression that quantify the linear relationship between variables.

The plot() function automatically selects the axis scales: on the x axis, we have values from $5.0e + 07$ or 50 million to $2.0e + 08$ or 200 million, with steps of 50 million.[2] On the y axis, we display 60–80 wins, at steps of 10 wins. We have the names of the variables as labels on the two axes. All elements of this plot are customizable.

As we mentioned above, the plot function can be used to draw multiple plots. It takes a type parameter that specifies the type of plot; by default, it is set to "p" or points. Other commonly used values of this parameter are "l" or lines, "b" or both points and lines.

2 By default, the plot() function uses the scientific notation to display large numbers. Here, 5.0e + 07 = 5.0×10^7 = 50,000,000.

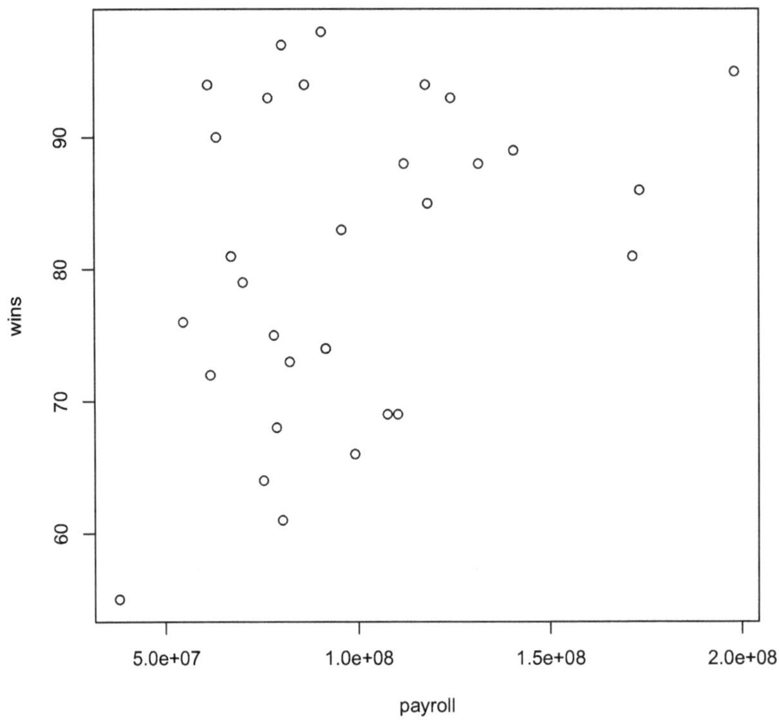

Fig. 4.1 A scatterplot between payroll and wins

4.2.1.1 Labeling Data Points

As the scatterplot displays all the data points of a dataset, it is a useful tool to visualize data trends. It is often useful to label a few data points in the scatterplot as examples. These could be the extreme points or cases of interest that we would like to highlight in our visualization. We use the identify() function for this purpose. We first create the plot using the plot() function as before.

We can find the data points that we want to label from the scatterplot itself. The identify() function waits for mouse clicks on the plot, as many specified by the n argument. The text displayed on the plot is given by the labels argument. Figure 4.2 shows the output of a scatterplot labeled using the identify() function.

```
> plot(payroll,wins)
> id = identify(payroll, wins, labels = code, n = 5)
```

In this plot, we label five data points using the `code` variable that contains the three-letter team name acronyms. From the data, we see that the team Washington Nationals (WSN) has the most wins (98). This is greater than many teams with much higher payrolls including the New York Yankees (NYY) and the Philadelphia

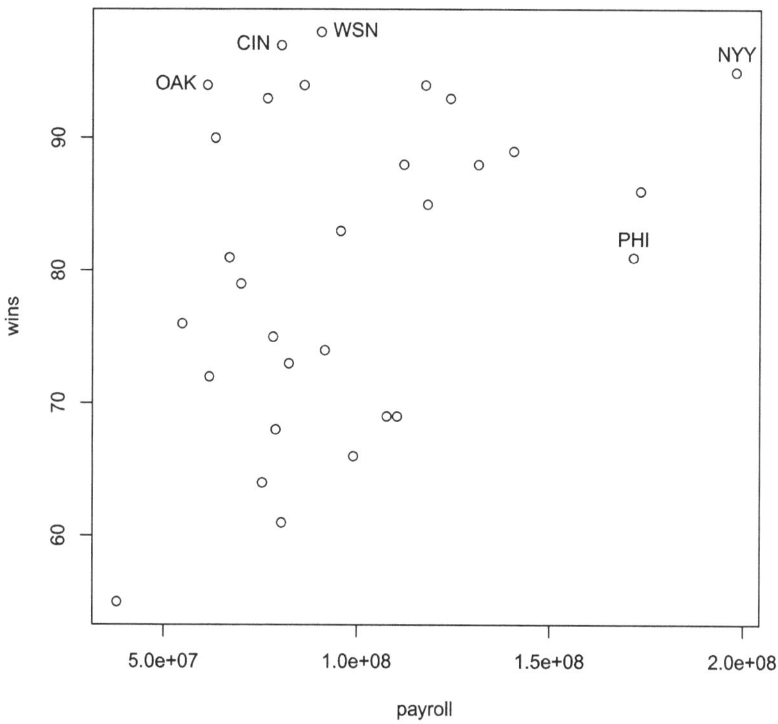

Fig. 4.2 A labeled scatterplot between payroll and wins

Phillies (PHI). A team that stands out is the Oakland Athletics (OAK) because of its consistent performance over the years, despite having a much lower payroll.[3]

Apart from labeling individual data points, we can display a piece of text on a plot using the text() function. This function takes the x and y coordinates, and the string that we want to display.

4.2.1.2 Points and Lines

In the above scatterplots, we displayed all the data points together. It is sometimes useful to denote different groups of data points by using different symbols or colors. This helps us in understanding if subsets of the data have different characteristics from the entire data set.

We create a scatterplot using data that are split on the league variable. We denote the teams belonging to the AL by triangles and the teams belonging to the NL by circles. One way to do so would be to first obtain the data points corresponding to either leagues, and then plot them separately. In the code below, we use the which()

[3] The success of the Oakland Athletics is the subject of the book "Moneyball: The Art of Winning an Unfair Game" [3], and the subsequent motion picture.

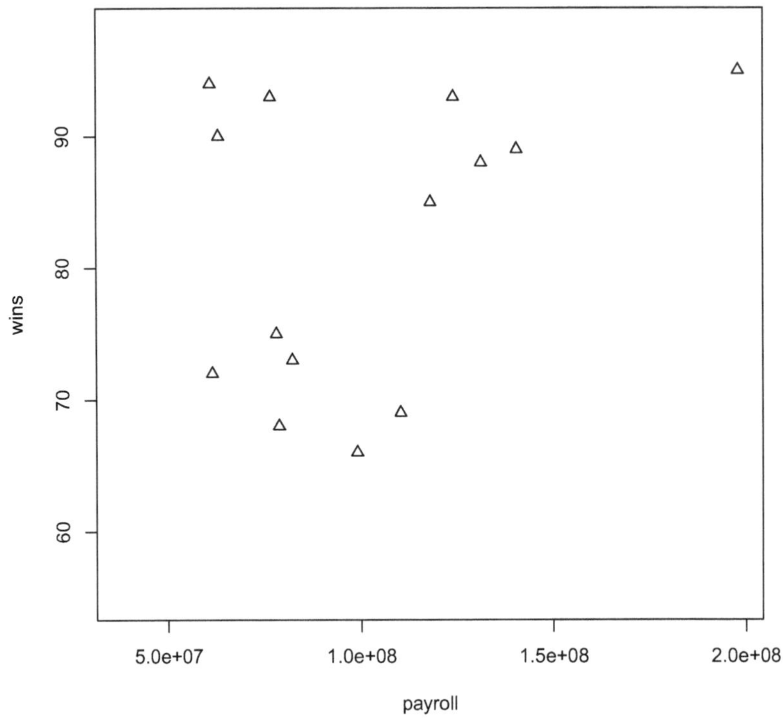

Fig. 4.3 A scatterplot between payroll and wins denoting American League (*AL*) teams by *triangles* and National League (*NL*) teams by *circles*

function to obtain two index vectors in the variables s1 and s2. We use the plot() function to plot the data points corresponding to NL alone, and then use the points() function to overlay the points corresponding to AL. Figure 4.3 shows the output.

```
> s1 = which(league == 'NL')
> s2 = which(league == 'AL')
> plot(payroll[s1],wins[s1],xlim=range(payroll),
        ylim=range(wins),xlab='payroll',ylab='wins')
> points(payroll[s2],wins[s2],pch=2)
```

The points() function takes the same as the plot() function: x and y coordinates of the data points. The only difference between the two is that points() does not regenerate the plot; it only adds a layer of new data points on top of the existing plot. We use the pch or plot character parameter to specify the shape that would denote the data points. There are 25 different shapes for pch as showed in Fig. 4.4.
In addition, we can assign characters to pch, e.g., `pch='x'` will plot the data points with the character `'x'`. Similarly, we can assign colors to the data points using the `col` parameter. This can be done using the numeric codes or literal names such as "red." We can obtain the list of color names using the colors() function. In the code above, we specified additional parameters to the plot function: the xlab and ylab

Fig. 4.4 Shapes for different pch values

parameters denote the labels for the x and y axes, and the xlim and ylim parameters denote their ranges. We did not need to specify xlim and ylim in the first example as the plot function uses the ranges of the input data by default; range(payroll) for the x axis and range(wins) for the y axis. Here, the problem is that we are using the subset payroll[s1] as the x variable, so the plot() function will use its range rather than the range of the entire payroll variable. This will omit the data points in payroll[s2] that fall outside this range, as the points() function does not redraw the axes. Using xlim = range(payroll) and ylim = range(wins) will ensure that all data points appear in the plot.

In R, there are usually multiple ways to do the same thing. We do not have to use the points() function to generate the scatterplot with data points split by a variable such as league. We can simply assign the variable to the pch parameter: by this, we will have a different type of plot character for different value of the splitting variable. We obtain the same plot as Fig. 4.3 in a more concise way below.

```
> plot(payroll,wins,pch=as.numeric(league),
       xlab='payroll',ylab='wins'))
```

The as.numeric() function converts a factor into a vector of integers, with a unique integer for each level.

```
> league[1:5]
[1] NL NL AL AL NL
Levels: AL NL
> as.numeric(league[1:5])
[1] 2 2 1 1 2
```

The technique of assigning values to the pch parameter is not restricted to factors. We can similarly draw a scatterplot with data points split on a numeric variable like pct. In the following code, we split the data points on the winning percentage or `pct >= 0.5`, i.e., the teams with half or more wins.

```
> plot(payroll,wins,pch=as.numeric(pct >= 0.5))
```

It is sometimes instructive to add lines to the plot that denote thresholds or trends. In this case, we draw a line at the `pct = 0.5` mark to separate the two groups of teams. The teams with `pct >= 0.5` will lie above this line and the teams with `pct < 0.5` will lie below this line. As all teams play 162 games, `pct = 0.5` equals 81 wins. We use the lines() function to draw lines overlaid on the scatterplot. This function takes the x coordinates and the y coordinates of the two end points. As we want to draw a horizontal line spread across the plot, the x coordinate of the end points is the range of the data, and the y coordinate of both end points is 81. Figure 4.5 shows the output.

```
> lines(range(payroll),c(81,81),lty=2)
```

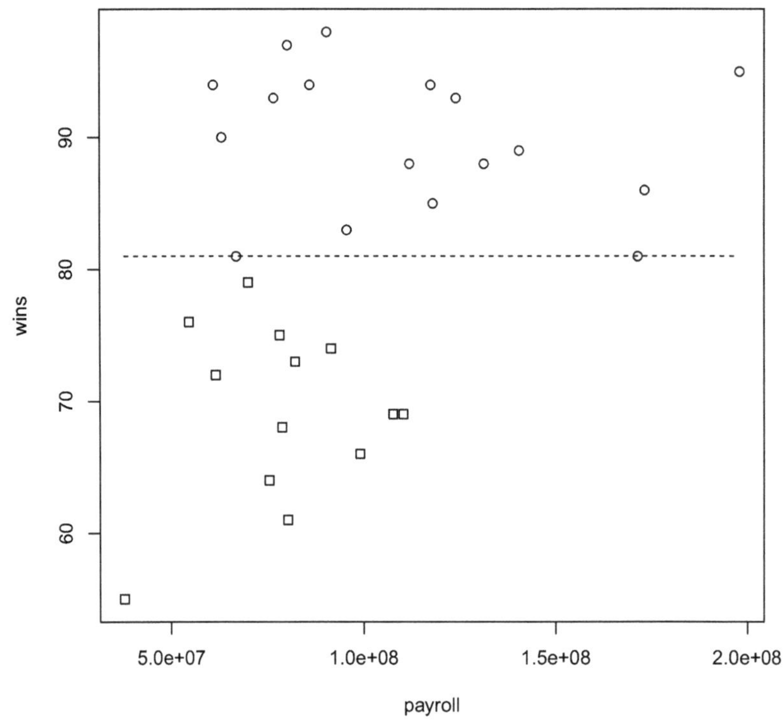

Fig. 4.5 A scatterplot between payroll and wins denoting teams with more than half wins by *circles* and less than half wins by *squares*

Fig. 4.6 Line styles for different lty values

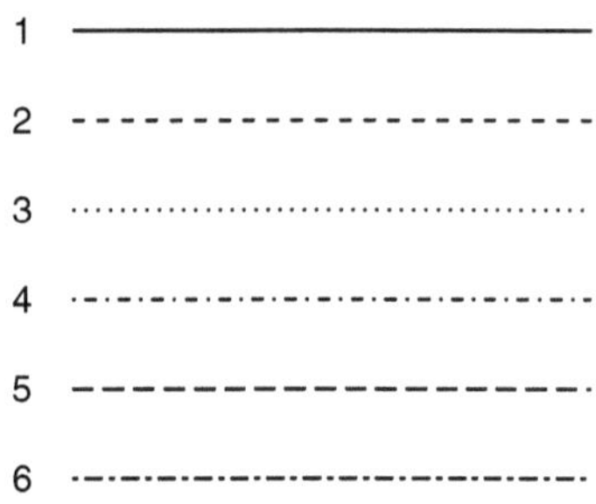

We set the lty parameter to 2 to draw a dashed line. Similar to pch, we can set the lty parameter to different values to draw lines of different styles as shown in Fig. 4.6.

If we have a scatterplot with multiple shapes, it is always useful to add a legend that denotes which shape corresponds to which grouping of the data. This helps a new observer in understanding what the plot is about. We add a legend using the legend() function that we call after the scatterplot is created. The legend() function takes as input the position where we would like to display the legend box, a vector of names corresponding to the data grouping, and the pch and col parameters we use to display

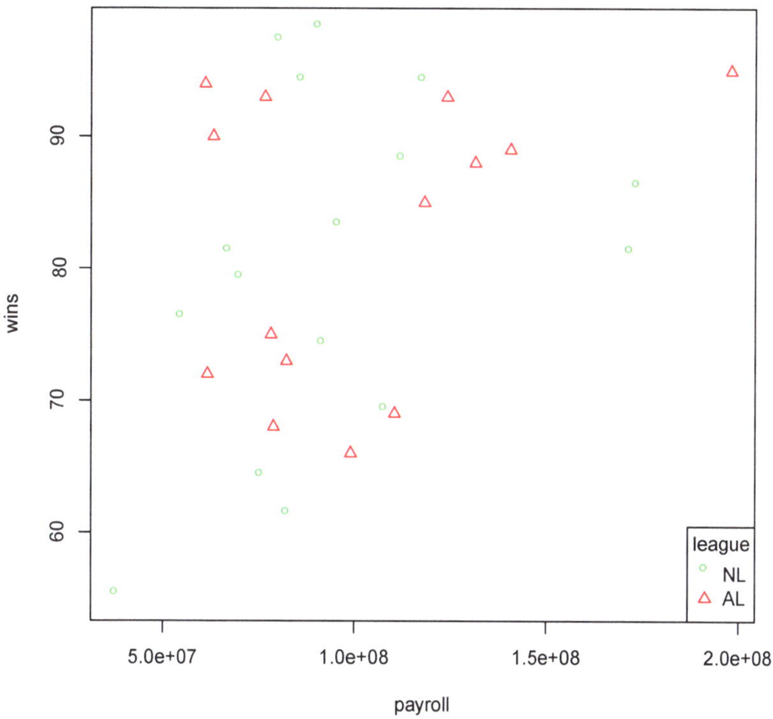

Fig. 4.7 A scatterplot with a legend

the data points in the plot. We can use the title parameter to display what the data grouping is. We can specify the location of the legend box in two ways: using x and y coordinates or a human-readable label, which is one of "bottomright," "bottom," "bottomleft," "left," "topleft," "top," "topright," "right," and "center." Figure 4.7 shows the output.

```
> legend('bottomright',c('NL','AL'),
        pch=c(1,2),title='leagues')
```

4.2.2 *Visualizing Aggregate Values with Bar plots and Pie charts*

Bar plot and pie charts are useful tools to visualize aggregated values of a quantitative variable. A bar plot contains a set of bars or rectangles, where the height of the bar corresponds to the value of the variable. Similarly, a pie chart visualizes the aggregated data as sectors of a circle.

We draw bar plots using the barplot() function and pie charts using pie() function. These functions take aggregated values as input. In R, there are multiple functions to aggregate data by a variable; here, we use the by() function.

4.2.2.1 Group Data by a Variable

Let us consider a case where we want to compute the total payrolls for teams at a division or league level. We can use the which() function to first identify the subsets of the data corresponding a group, e.g., `which(league == 'NL')`, and then apply the aggregation function to this group:

```
> sum(payroll[which(league == 'NL')])
  [1] 1512099665
```

To obtain the total payroll for teams at each aggregation level, we will need to do the above steps with all values of the league and division variables. This may suffice for variables with a couple of values but it is cumbersome for multivalued variables.

The by() function allows us to perform operations on subsets of data that are split by different values of a variable. This function takes as input the input data, the variable on which we want to split the data, and the function we want to apply to the subsets. Here, the variable is analogous to the GROUP BY clause of a SELECT query in SQL. If we want to compute the total payroll of teams at a league level, we can use the by() function as follows:

```
> by(payroll,league,sum)
league: AL [1] 1424254675
-------------------------------------------------------
league: NL [1] 1512099665
```

The by() function takes three inputs: the variable that we need to aggregate, the variable we want to group by, and the aggregation function. We can also use the by() function to aggregate on multiple variables by passing a list of variables as the second input.

```
> by(payroll,list(division,league),sum)
: Central
: AL
:[1] 489326375
-------------------------------------------------------
: East
: AL
:[1] 530789300
-------------------------------------------------------
: West
: AL
:[1] 404139000
-------------------------------------------------------
: Central
: NL
:[1] 476248200
-------------------------------------------------------
: East
: NL
```

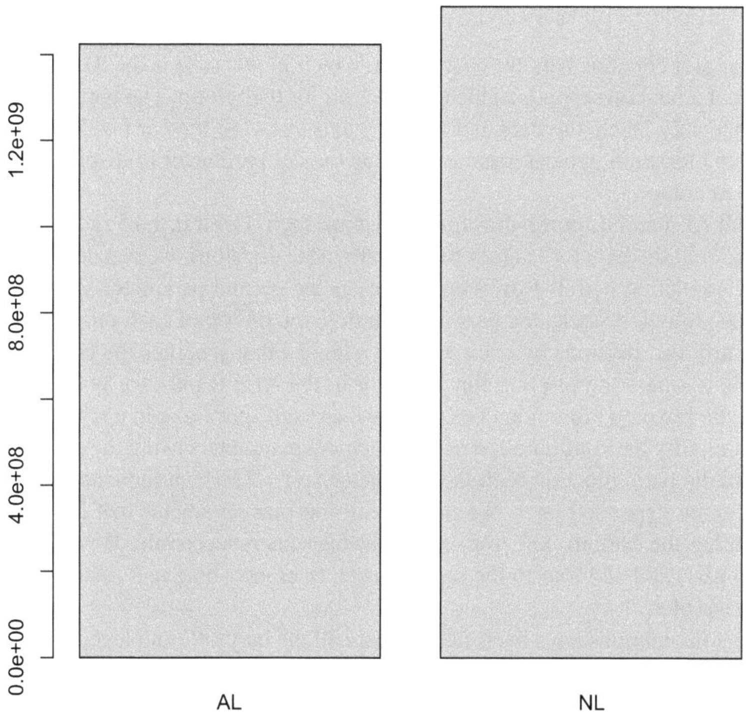

Fig. 4.8 Bar plot comparing total payrolls of American League (*AL*) and National League (*NL*) teams

```
:[1] 547594982
--------------------------------------------------------
: West
: NL
:[1] 488256483
```

The by() function is an example of a higher order function as it takes the aggregation function as an argument. After splitting the data into different groups, by() applies the function specified by the third argument to each group, without even knowing what that function will be ahead of time.

Internally, by() is a wrapper around the tapply() function, which also provides similar output. Other useful higher order functions in R include sapply() and mapply(), which apply a function to each elements of a vector. There are also popular packages such as pylr and funprog that have more sophisticated functions for aggregating data.

4.2.2.2 Bar Plots

The barplot() function consumes the output of the by() function to create the bar plot with the aggregated data. Figure 4.8 shows the output of total payroll for teams at a league level.

```
> barplot(by(payroll,league,sum))
```

The bar plot contains bars for each group, which in this case is the AL and NL. The height of a bar corresponds to the total payroll for that group. The barplot() function automatically labels the bars and adds a y axis showing the bar heights. Similar to the plot() function, we can pass a vector as the col parameter to display the bars in different colors.

Both AL and NL consist of three divisions each: Central, East, and West. If we want to split the bars for leagues further into these divisions, we can do so by calling the by() function with list(division,league) as the second parameter. By default, the barplot() function stacks the bars for each division on top of each other, so we need to identify the divisions by color and add a legend that specifies the labeling.

This is where the plotting functionality in the base R package gets a bit tricky; often, the boxplot() function does not leave enough space to add a legend. We need to first modify the configuration of the graphical parameters using the par() function. We call the par() function with the parameter xpd = T, which indicates that the plot area can be expanded later. We then tweak the mar parameter that holds a vector containing the bottom, left, top, and right margins respectively. We make another call to par() and add four to the right margin, to create some real estate on the right of the bar plot.

After the margins have been set, we first call the barplot() and legend() functions. We need to manually set the position of the legend box using the x and y coordinates. By default, the barplot() function stacks the bars in an alphabetical order of the grouping variable, which in this case is division. Figure 4.9 shows the bar plot that we obtain using the code below.

```
> par(xpd=T, mar=par()$mar + c(0,0,0,4))
> barplot(by(payroll,list(division,league),sum),
  col=2:4)
> legend(2.5,8e8,c('Central','East','West'), fill=2:4)
```

Having stacked bars makes it difficult to compare the values for different divisions. To display the bars besides each other, we need to call barplot() with the parameter `beside = T`. Figure 4.10 shows the output.

```
> par(xpd=T, mar=par()$mar + c(0,0,0,4))
> barplot(by(payroll,list(division,league),sum),col=
  2:4,beside=T)
> legend(8.5,3e8,c('Central','East','West'),fill=2:4)
```

4.2.2.3 Pie Charts

A pie chart is an alternative visualization that shows the aggregated value of a variable computed over different subgroups of the data. The aggregated values for different subgroups are shown as a portion of the whole. Here, the circle represents the

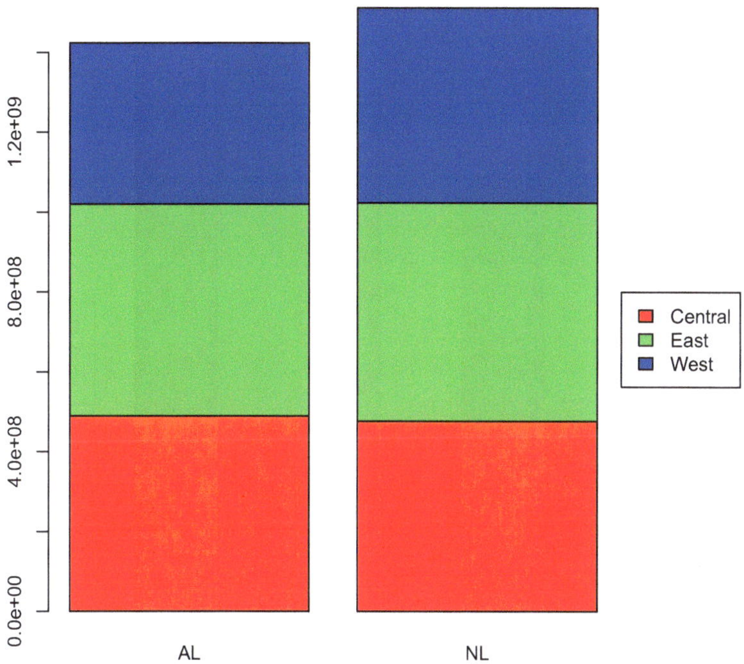

Fig. 4.9 Bar plot comparing total payrolls of American League (*AL*) and National League (*NL*) teams

aggregated value computed over the entire dataset, and slices or sectors represent those computed over different data subgroups.[4]

A pie chart conveys exactly the same information as a bar plot: both provide aggregated values for a variable over different data subgroups. Also, both the charts use geometric area to indicate this: rectangles for bar plots and sectors for pie charts. In addition, there is an advantage to bar plots as they have a scale on the y axis that allows us to easily compare the heights of different bars. It is more difficult to visually compare the areas of slices, especially across different pie charts. In statistics, the use of pie charts has been discouraged for this reason [2; 5]. Nevertheless, pie charts remain widely used in business and popular media.

We draw a pie chart using the pie() function. Similar to barplot(), this function also consumes aggregated data that is the output of the by() function. We create a pie chart of total payrolls of AL and NL teams below.[5] Figure 4.11 shows the output.

```
> pie(by(as.numeric(payroll), league, sum))
```

[4] It is called a pie chart because of its resemblance to slices of a pie.

[5] We use the as.numeric() function to prevent an overflow as sum(payroll) has values larger than the pie() function can handle. For aggregate variables with smaller values, calling as.numeric() is not necessary.

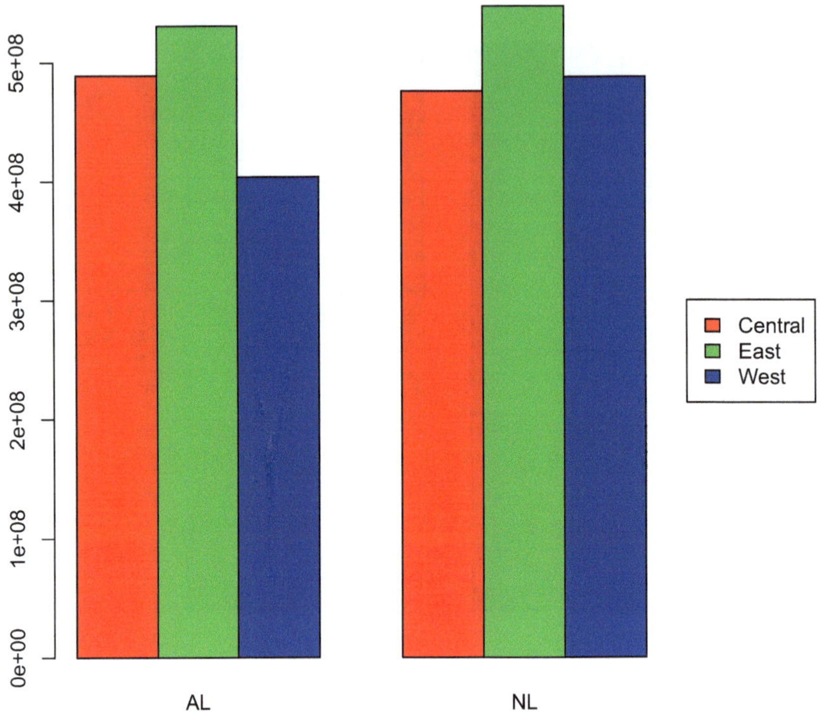

Fig. 4.10 Bar plot comparing total payrolls of American League (*AL*) and National League (*NL*) teams

The pie chart consists of a circle with two slices, a shaded one corresponding to the NL and an unfilled one corresponding to the AL. The slice for NL is larger, implying the total payroll for NL teams is greater than that of the AL teams. This is consistent with the bar plot that we had in Fig. 4.8.

We can also use the pie() function to draw a pie chart of the payroll of teams split by division and league by simply calling the by() function to group on these two variables. In this case, the pie() function does not automatically label the slices, so we need to call it with a vector of labels ourselves.

```
> labels = c('AL Central', 'AL East', 'AL West',
             'NL Central', 'NL East', 'NL West')
> pie(as.numeric(by(payroll,list(division,league),
  sum)),labels)
```

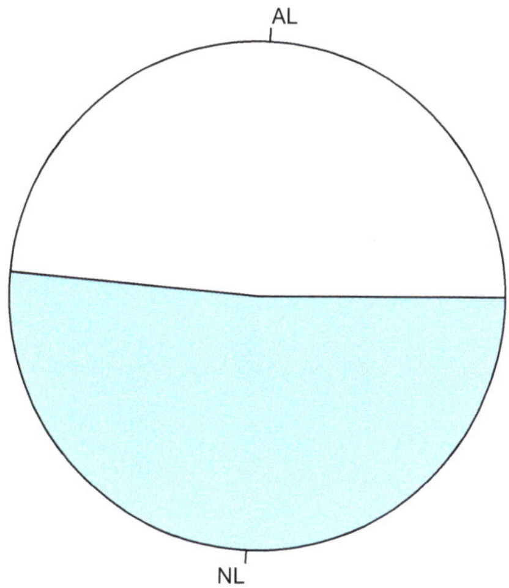

Fig. 4.11 Pie chart comparing total payrolls of American League (*AL*) and National League (*NL*) teams

4.2.3 Common Plotting Tasks

4.2.3.1 Multiple Plots

We often need to display multiple visualizations together. This could be the same type of plot drawn on different subsets of the dataset or different type of plots drawn on the same dataset. The mfrow parameter of the par() function allows us to draw multiple plots in a grid. This parameter takes a vector of two elements that denote the number of rows and columns of the plot.

In the code below, we draw two plots side by side by setting mfrow to one row and two columns: c(1,2). As an example, we then draw separate scatterplots for payroll versus wins for NL and AL teams. We label the plots by setting a title using the main parameter of the plot() function. Figure 4.13 shows the output.

```
> par(mfrow=c(1,2))
> s1 = which(league == 'NL')
> s2 = which(league == 'AL')
> plot(payroll[s1],wins[s1],main='NL',xlab='payroll',
  ylab='wins')
> plot(payroll[s2],wins[s2],main='AL',xlab='payroll',
  ylab='wins')
```

In general, we can display any visualization in the grid including those generated by the barplot() and pie() functions. We can continue to call the plot() function to replace the plots displayed in the grid. The lifetime of the grid is till the plot window is open; after that, we need to call the par() function with the mfrow parameter again.

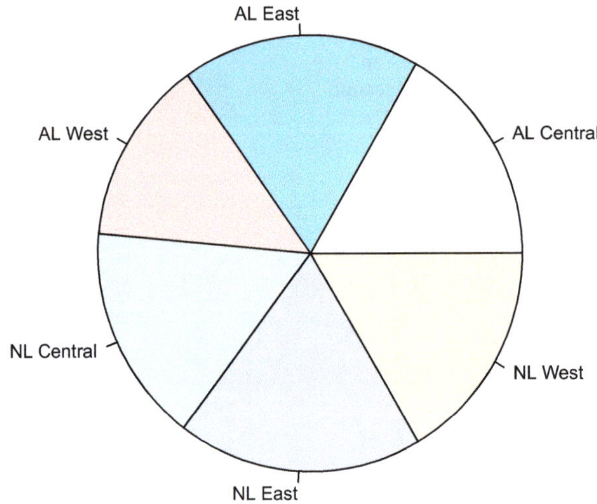

Fig. 4.12 Pie chart comparing total payrolls of teams grouped by league and division

4.2.3.2 Saving Plots to Files

When the plot() function is invoked, it draws the plot on the onscreen display. The lifetime of this plot is only till the window is open. In most data analysis tasks, we would also need to persist the plot somewhere, so that we can use it later in a presentation, report, or a book. R has support for saving the plot in multiple file formats. Using this functionality is not very intuitive for beginners, but it is fairly simple.

R has a concept of a graphics device, which is conceptually similar to a physical output peripheral like a display or a printer. When the plot() function is invoked, R sends the data corresponding to the plot over, and the graphics device generates the plot. The default graphics driver is X11 that displays the plot in a window frame. To save the plot to a file, we need to use a graphics device corresponding to the file format we are interested in. For most plots, it is recommended to use lossless file formats like PNG or PDF.[6] Table 4.2 has the list of available graphics devices.

To save a plot to a file format such as PDF, we first open a device using the pdf() function called with the output file name. After we are done, we call the the dev.off() function that closes the device and completes writing the plot to the file. It then sets the display device back to the default, which will display the next plot in a window. We write one of our plots to a file "plot.pdf" below.

```
> pdf('plot.pdf')
> plot(payroll,wins)
> dev.off()
```

The above mechanism is also useful when we have a script that automatically generates the plots. If the plot is already displayed in a window, an alternative way to

[6] All plots appearing in this book were created using the pdf graphics device.

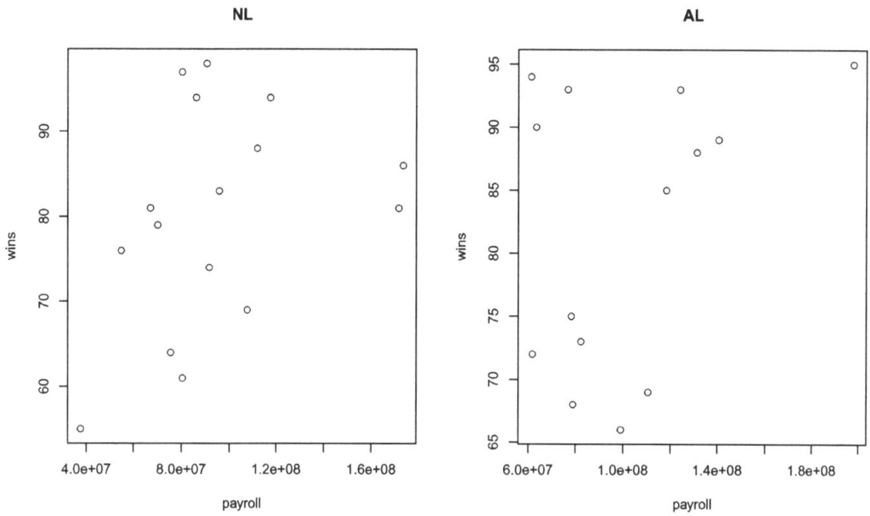

Fig. 4.13 Multiple plots

Table 4.2 List of graphics devices

Device	Description
pdf	PDF file format
postscript	PostScript file format
xfig	XFIG graphics file format
bitmap	bitmap pseudodevice via Ghostscript
X11	X11 windowing system
svg	SVG file format
png	PNG file format
jpeg	JPEG file format
bmp	BMP file format
tiff	TIFF file format

save it to a file is using the dev.copy() function. As above, there are several variants of this function for each file format. We use the one for PDF below; by default, this function saves the plot to the file Rplot.pdf.

```
> plot(payroll,wins)
> dev.copy2pdf()
```

4.3 Layered Visualizations Using ggplot2

The plotting functions that we discussed in the previous section are from the R base package. This is not the only way to obtain visualizations in R. Other popular visualization packages include lattice and ggplot2. The ggplot2 package [6] stands out because of its conceptually different approach of dealing with visualizations.

The ggplot2 visualization package is based on the theory of grammar of graphics [7] and derives its name from the same theory. In this framework, a visualization is divided into four components or layers:

1. Data to be visualized
2. Aesthetics of the visualization: e.g., color, shape, and scale
3. Geometric objects: e.g., points, lines, and polygons
4. Statistical transformation: e.g., smoothing and binning

The ggplot2 package allows us to specify each of these components separately. This makes it convenient to create complex and multilayered visualizations. We can even create different visualizations by replacing only some of the visualization components. Also, the ggplot2 framework takes care of the supporting features such as creating the legend.

This concept of layers is different from the base graphics package where we need to specify all the details of the visualization in the plot() function including positions of individual elements. There is nothing inadequate about the base graphics package; we can use it to generate the same visualizations that we can generate using ggplot2. The latter gains from its modular framework when we want to create complex visualizations with multiple layers. In the base graphics package, we usually need to configure a few parameters to create visualizations of the same aesthetic appeal. In general, the visualizations created using the default settings of ggplot2 tend to look nicer. The only downside of ggplot2 is that it is not designed for efficiency: it is many times slower than the base graphics package. This is not an issue with small-to-medium-sized datasets, but can be one when dealing with large datasets.

The ggplot2 package has two functions: `qplot()` and `ggplot()` that we use to create the visualizations. The `qplot()` function has a similar interface to `plot()`, while the `ggplot()` function provides the full grammar of graphics interface. We first look at the `qplot()` function.

4.3.1 Creating Plots Using qplot()

Just like the plot() function, we can use the `qplot()` function to draw a scatterplot between two variables. We draw a scatterplot between team payrolls and wins below. Figure 4.14 shows the output. As the `ggplot2` package is not part of the base R distribution, we first need to include it by calling library(ggplot2).

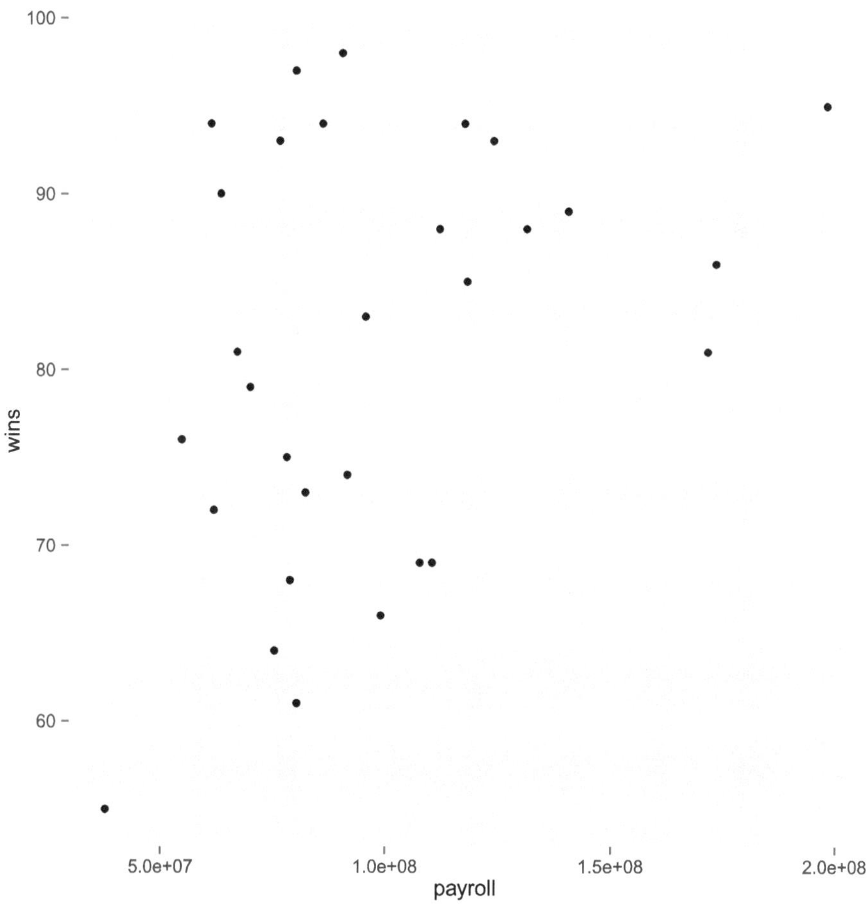

Fig. 4.14 Scatterplot between team payrolls and wins generated by qplot()

```
> library(ggplot2)
> qplot(payroll, wins)
```

This scatterplot has the same visualization that we had in Fig. 4.1; `qplot()` adds a gray background and a grid by default.

`qplot()` has many parameters that we can use to configure the plot. The `col` and `shape` parameters control the color and geometric shape that are used to denote the points in the scatterplot. To identify the teams by their leagues and divisions, we set these parameters with the league and division variables. Figure 4.15 shows the output.

```
> qplot(payroll,wins,col=league,shape=division)
```

In the visualization, the AL and NL teams are denoted using red and blue colors respectively, and the teams from Central, East, and West divisions are denoted using

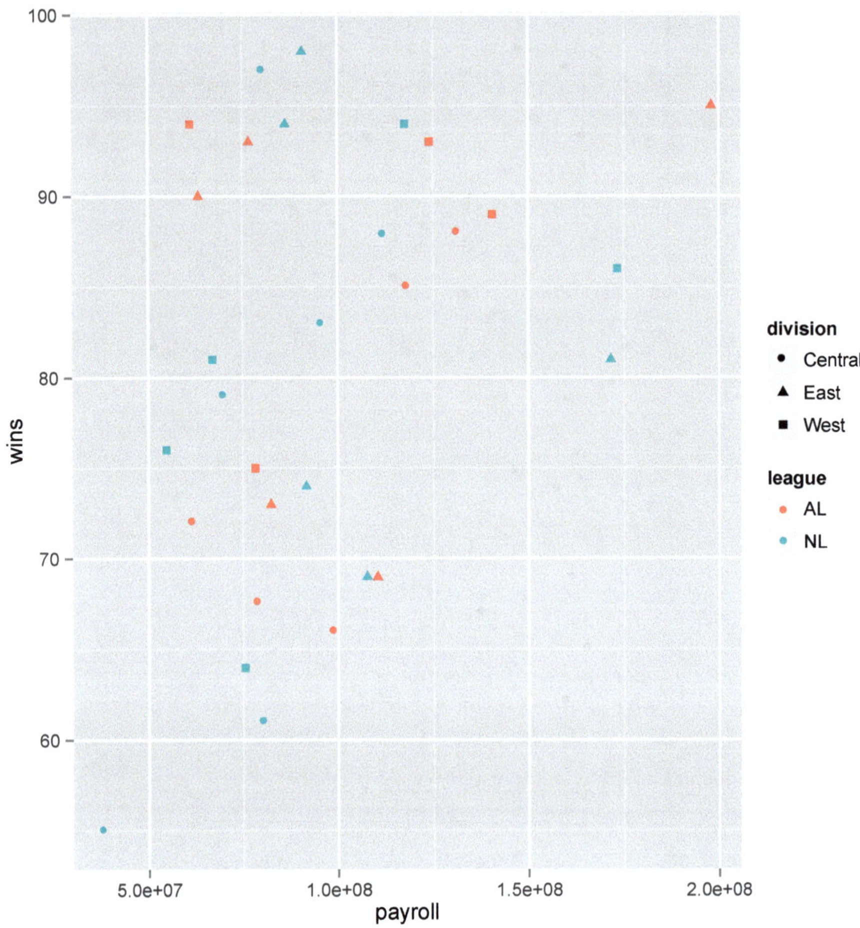

Fig. 4.15 Scatterplot with leagues and divisions

circles, triangles, and squares, respectively. We can identify all the six combinations of league and divisions using these two colors and three shapes. `qplot()` also automatically adds a legend to the right. This is convenient considering the extra work we needed to add a legend manually using the legend() function.

We can use the `qplot()` function to draw bar plots as well. We need to specify two things: the variable that we want to group the data as the input data, and the variable we want to aggregate as the weight parameter. The aggregation functionality is built into qplot(); we do not need to use the by() function as we did with the plot() function. We draw a bar plot for total payroll for teams per league below. Figure 4.16 shows the output.

```
> qplot(league,weight=payroll,ylab='payroll')
```

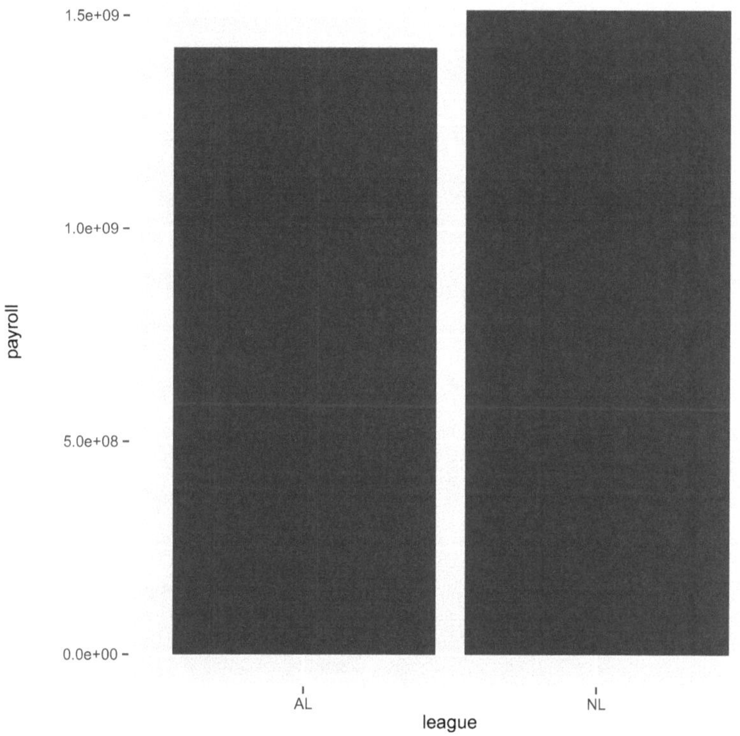

Fig. 4.16 Bar plot of total payrolls of the two leagues

`qplot()` has a fill parameter that we can use to specify the color of the bars. We can create a stacked bar plot by setting a variable to the fill parameter, which is similar to setting the col and shape parameter that we did above. We split the bars further by division below. Figure 4.17 shows the output.

```
> qplot(league,weight=payroll,ylab='payroll',
  fill=division)
```

Similar to the scatterplot, the `qplot()` function automatically adds a legend identifying the divisions by fill color.

The layout of the stacked bars is controlled by the `position` parameter of the `qplot()` function. We can stack the bars side by side by setting it to `'dodge'`. Figure 4.18 shows the output.

```
> qplot(league,weight=payroll,ylab='payroll',
        fill=division,position='dodge')
```

`qplot()` can also be used to generate pie charts by changing bar plots to polar coordinates.

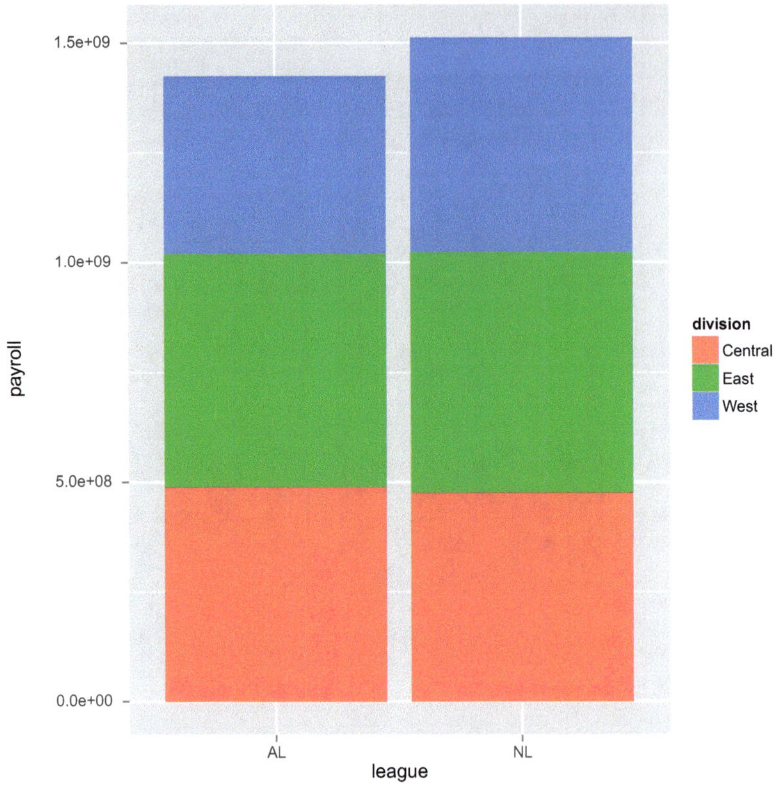

Fig. 4.17 Bar plot of total payrolls of the two leagues split by division

Another advantage of `qplot()` is that we do not need to use par() to display variants of a visualization in a grid. We use the facet parameter of the `qplot()` function to specify a formula containing variables with which we want to vary the visualization. We create bar plots for the total payroll per division for the two leagues using the formula `. ~ league`.[7] Figure 4.19 shows the output. There is no need for a legend because all division and league combinations are already labeled.

```
> qplot(division,weight=payroll,ylab='payroll',
  facets= . ~ league)
```

The facet parameter works for other visualizations generated by the qplot() function including scatterplots.

[7] Formulae are first class objects in R. We will look at them more closely in the following chapters.

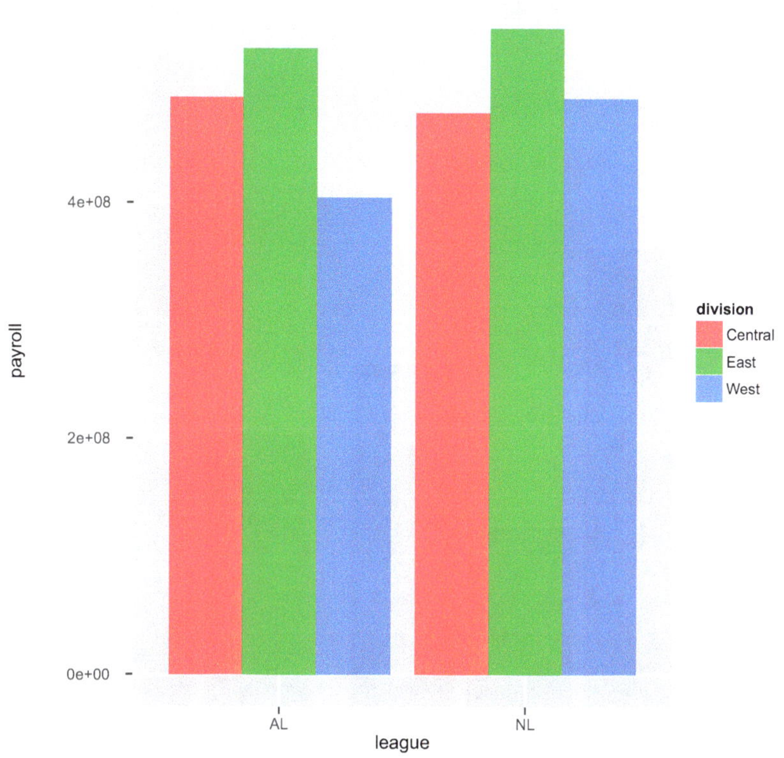

Fig. 4.18 Bar plot of total payrolls of the two leagues split by division, placed side by side

4.3.2 ggplot(): Specifying the Grammar of the Visualization

As we discussed above, the `ggplot` package is based on the grammar of graphics. Using the ggplot() function, we can specify individual elements of the visualization separately. The syntax for `ggplot()` is different from that of `qplot()`, although the same visualizations can be generated by both functions.

The basic unit of a visualization is the ggplot object. It is the abstract representation of the plot object and only contains the data frame that we are interested in visualizing. We create this object using the `ggplot()` function. This does not create any visualization by itself.

```
> p = ggplot(data)
```

To create a visualization, we need to add other plot elements such as aesthetics and layers. The aesthetics include the variables, geometric shapes, and colors that we want to display, and layers contain information about how to display them. We create a scatterplot between team payroll and wins below by specifying these variables using

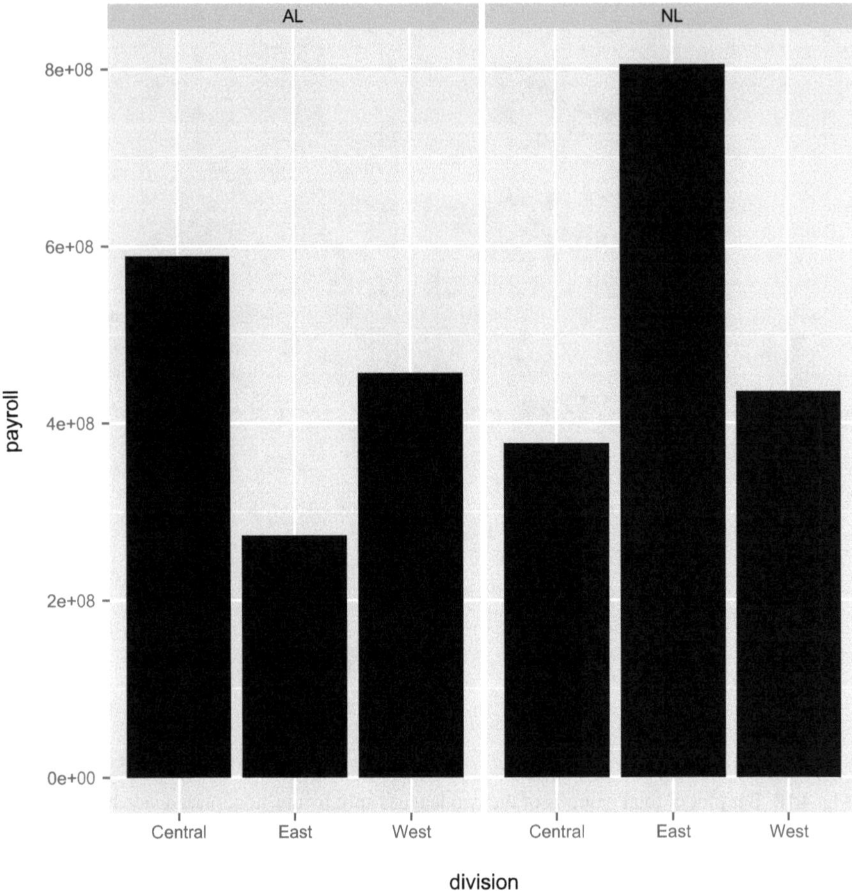

Fig. 4.19 Bar plot of total payrolls of the three divisions per division split by league

the aes() function and adding a points layer using the layers() function. The output is identical to Fig. 4.15 we obtained using `qplot()`.

```
> p + aes(payroll,wins,shape=division,col=league)
  + layer('point')
```

This syntax is a bit esoteric: we use the + operator to add components to the ggplot object. We can also create a grid by adding the `facet_grid()` function.

We create bar plots by specifying the layer as bar. We also need to set the variable that we want to display as the weight parameter in the aesthetics component. The ylab() function specifies the label of the y axis. The output is identical to Fig. 4.17.

```
> p + aes(league,weight=payroll,fill=division)
  + ylab('payroll') + layer('bar',position='dodge')
```

We reuse the same ggplot object `p` for drawing both the scatter and bar plot.

4.3.3 *Themes*

Although the default theme of ggplot suffices for most purposes, sometimes we might want to customize our visualization to a particular color and font combination. A bonus feature of the ggplot package is that it comes built-in with a set of visualization themes through the `ggthemes` package.

There are many themes available in the `ggthemes` package, including those used in popular publications such as the Economist and the Wall Street Journal.[8] To apply a theme, we need to use the appropriate theme function as a layer to the ggplot object. As an example, we add The Economist theme in the code below. Figure 4.20 shows the output.

```
> library(ggthemes)
> p + aes(payroll,wins,shape=division,col=league)
  + layer('point') + theme_economist()
```

The output visualization contains a scatterplot with a blue background similar to a visualization appearing in The Economist magazine.

4.4 Interactive Visualizations Using Shiny

In all of the techniques that we discussed so far, the visualization output is in the form of a static image that can be displayed on screen or saved as an image file. To allow users to explore the data analysis on their own, it is useful to create interactive visualizations. This requires creation of a user interface (UI) that captures the user inputs. There are a few packages that allow us to create interactive visualizations in R such as rggobi, iplot, gWidgetsWWW2, and `shiny`. In this section, we look at creating a Web application in `shiny`.

`Shiny` is a more recent R package that allows us to create interactive UIs in the form of Web applications. A major benefit of creating Web applications is that a large number of users can access the interactive visualization through their Web browsers without requiring to install any software package including R and downloading the data.

One advantage of `shiny` is that we can create rich Web applications entirely in R and without writing any HTML or Javascript code. Although, it is possible to specify the UI using an HTML page, if needed; in this section, we will look at creating a Web application in R itself. A `shiny` application consists of two R scripts: ui.r and server.r. The ui.r script contains a specification of the UI elements including the layout of the controls and the output. The server.r script contains the application logic that generates the plot.

[8] A gallery of themes is available at https://github.com/jrnold/ggthemes.

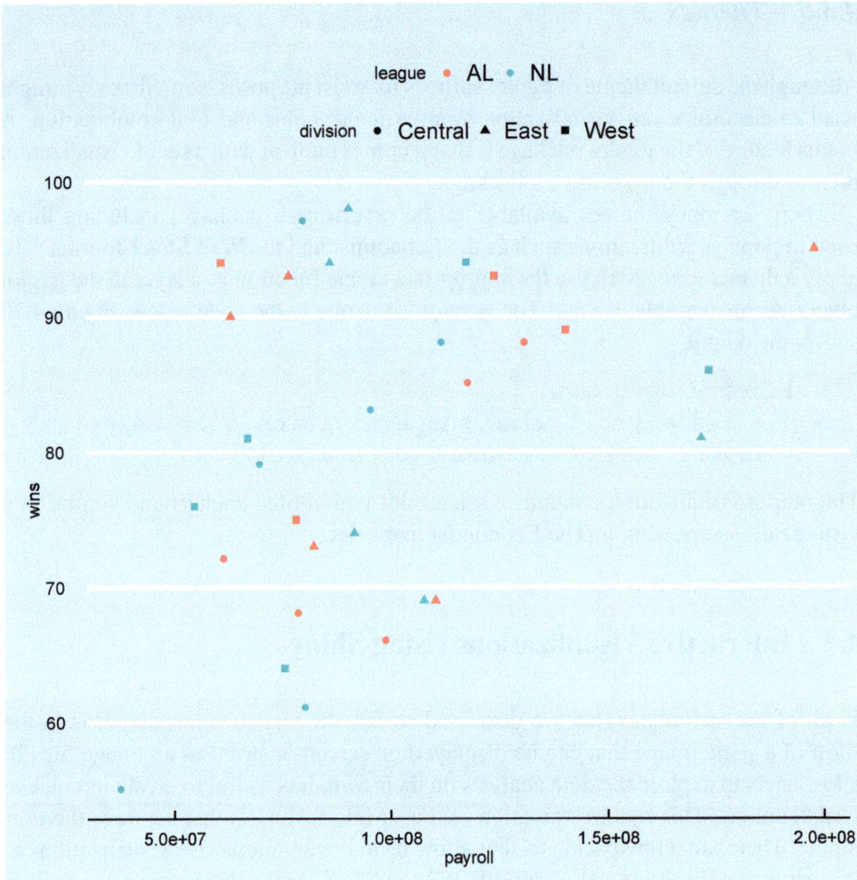

Fig. 4.20 Scatterplot between team payroll and wins with The Economist theme

In the discussion below, we only scratch the surface of creating an interactive visualization; `shiny` has a fairly rich feature set that we can use to create complex Web applications. The official tutorial [4] is a good place to learn more about the advanced features of `shiny`.

As an example, we create an interactive visualization to allow users to access the MLB dataset through a scatterplot between team payroll and wins. We allow the users to identify the data points in the scatterplot by league and division. To do so, we need two UI elements: we choose to go with two check boxes that group the data points by league and division, respectively.

We are interested in creating an interactive visualization as shown in Fig. 4.21. We specify the UI in the ui.r script below. This script contains one function `shinyUI()` and we need to add UI elements as its parameters. We define a page with sidebar in order to have the controls in the left panel and the plot in the remainder of the window.

Fig. 4.21 A shiny Web application

We specify the header of the page using the `headerPanel()` function, and the sidebar and the main panels using the `sidebarPanel()` and `mainPanel()` functions, respectively. In each of these functions, we specify the controls for the respective panels; the two check boxes: `league` and `division` for the sidebar and the plot element `payroll.wins` itself for the main panel. This script is intended to run independently and so, it first needs to load the `shiny` package.

```
# ui.r

library(shiny)

shinyUI(pageWithSidebar(
    headerPanel('payroll vs wins'),

    sidebarPanel(
        helpText('Group by'),
        checkboxInput('league', 'League', FALSE),
        checkboxInput('division', 'Division', FALSE)
    ),
```

```
        mainPanel(
            plotOutput('payroll.wins')
        )
    ))
```

Apart from check boxes, shiny supports a wide variety of UI elements including text boxes, drop-down lists, and sliders.

The purpose of the server.r script is to access the user inputs provided by the UI and generate the appropriate plots. This script contains the shinyServer() function that takes another anonymous function with UI inputs and outputs as its input. The inputs and outputs are available as the input and output data frames; we can access the checkboxes through input$league and input$division variables, and we need to assign the plot object to the output$payroll.wins variable. We create the plot using the renderPlot() function. We load our data set separately using the read.csv() function.

Although any R visualization package can be used to create plots for shiny applications, we use the ggplot2 package here. There are certain advantages of this choice; as we shall see later, the modular interface of ggplot allows us to add layers for the visualization in response to the user input. In the renderPlot() function, we first create a ggplot object, and add payroll and wins as variables, and a scatterplot. As we have seen in the previous section, this creates a scatterplot with payroll and wins, with none of the data points highlighted. If the check boxes for league and division are set to true, we add additional layers to the ggplot object for highlighting the data points by color and shape. To display the ggplot object as a scatterplot, we finally use the print() function.

```
# server.r

library(shiny)
library(ggplot2)

data = read.csv('teams.csv')

shinyServer(function(input,output) {
  output$payroll.wins = renderPlot({
    p = ggplot(data)
    p = p + aes(payroll,wins) + layer('point')

    if (input$league == 'TRUE') {
      p = p + aes(col=league)
    }

    if (input$division == 'TRUE') {
      p = p + aes(shape=division)
    }
```

```
    print(p)
  })
})
```

The Web applications created using shiny are reactive: the plot output changes immediately in response to changing the user input, without reloading the Web page. Internally, this works by calling the shinyServer() function each time there is a change in the user input. We do not need to write custom code for this; the shiny Web framework handles the user events automatically.

The shiny package makes it simple to run and test the Web application. We only need to invoke the runApp() function with the path of the directory containing the ui.r and server.r scripts. We run the Web application located in the shiny-app/directory below.

```
> runApp('shiny-app/')
  Listening on port 8100
```

This function starts the shiny Web server inside the R process. The Web application is available at http://localhost:8100/ by default. The runApp() function also opens the default Web browser with the Web application automatically; otherwise, we need to open a Web browser and point it to this URL.

The above method is adequate for developing and testing Web applications. The shiny-server package[9] is more appropriate for running shiny applications in production.

4.5 Chapter Summary and Further Reading

In this chapter, we looked at different ways of visualizing the data in R. Some of the commonly used visualizations include scatterplots, bar plots, and pie charts. Using the R base package, we can create these visualizations using the plot(), barplot(), and pie() functions. The base package is usually adequate for creating simple visualizations but it can become cumbersome when creating the more complex ones. The ggplot2 package allows for creating complex visualizations easily because of its multilayered design, with each layer representing an aspect of the visualization. The ggplot2 package provides two functions, qplot() and ggplot(), to create the visualizations. The visualizations mentioned so far are mostly static; the shiny package allows us to display our visualization as an interactive Web application that can change based on the user inputs.

The visualizations we saw in this chapter are for datasets with relatively small number of data points and variables. The ggobi package supports creating visualizations for high-dimensional datasets. The bigvis package allows for creating

[9] https://github.com/rstudio/shiny-server.

visualizations for large datasets up to 100 million data points in a few seconds. We can also visualize nonrelational data in R: the igraph package allows for visualizing network structures such as a social network. The `maps` and `ggmaps` packages can be used to create visualizations of geographical data superimposed on a map of an area.

References

1. http://blog.revolutionanalytics.com/2011/03/how-the-new-york-times-uses-r-for-data-visualization.html
2. http://www.stevefenton.co.uk/Content/Pie-Charts-Are-Bad/.
3. Lewis, M. (2004). *Moneyball: The art of winning an unfair game*. New York: W. W. Norton & Company.
4. Teach yourself shiny. http://shiny.rstudio.com/tutorial/.
5. Tufte, E. (2001). *The visual display of quantitative information*. Cheshire: Graphics Press.
6. Wickham, H. (2009). *ggplot2: Elegant graphics for data analysis*. Use R!. New York: Springer.
7. Wilkinson, L. (2005). *The grammar of graphics*. New York:Springer.

Chapter 5
Exploratory Data Analysis

When we are analyzing a new dataset, it is beneficial to get a sense of the layout of the data first. This layout is in the form of how the data is structured or more concretely, the data distribution. Exploratory data analysis (EDA) is a collection of analysis techniques that we can apply to the data for this purpose. Most of these techniques are often simple to implement as well as computationally inexpensive, which allow us to obtain the exploratory results quickly.

EDA helps us form our intuitions about the underlying processes that explain the data. This in turn helps us in choosing which analysis method to apply to the dataset. EDA is an important step in the life cycle of a data science project. In many such projects, which data science methodology to use is often unclear and so is the potential that the dataset has to offer. A good starting point is to look around the dataset to see if we can find something interesting; we can form detailed analysis plans after that.

In this chapter, we will look at various ways to get an overview of the data. This includes summary statistics like maximum and minimum values, means, medians, and quantiles. We also look estimating the data distribution by using box plots and histograms. To get a numeric measure of the symmetry of the data distribution, we also look at advanced measures such as skewness and kurtosis. Finally, problem of using exploratory analysis techniques to find outliers in the data.

Case Study: US Census

To illustrate exploratory data analysis, we use data from an important dataset: the 2010 US Census.[1] The dataset contains the variables listed in Table 5.1. The areas are classified into counties or equivalent, metropolitan areas, metropolitan statistical areas, and micropolitan statistical areas.[2] In the dataset, the type of an area is denoted by the legal/statistical area description (LSAD) variable.

---

[1] Obtained from the US Census Bureau http://www.census.gov/popest/data/counties/asrh/2011/CC-EST2011-alldata.html.

[2] These terms are defined by the Office of Management and Budget of the US federal government. http://www.whitehouse.gov/sites/default/files/omb/assets/fedreg_2010/06282010_metro_standards-Complete.pdf.

M. A. Pathak, *Beginning Data Science with R*,
DOI 10.1007/978-3-319-12066-9_5

Table 5.1 List of variables in census data

Variable name	Description
NAME	Name/title of area
LSAD	Legal/Statistical Area Description
CENSUS2010POP	2010 Resident Census population
NPOPCHG_2010	Numeric change in resident total population
NATURALINC2010	Natural increase in the period (4/1/2010–7/1/2010)
BIRTHS2010	Births in the period
DEATHS2010	Deaths in the period
NETMIG2010	Net migration in the period
INTERNATIONALMIG2010	Net international migration in the period
DOMESTICMIG2010	Net domestic migration in the period
RESIDUAL2010	Residual for the period

We can load the dataset using `read.csv()`:

```
> data = read.csv('metropolitan.csv')
```

5.1 Summary Statistics

Summary statistics is an integral part of data exploration. We first look at how to compute summary statistics in *R* using the data from the case study.

5.1.1 Dataset Size

The first statistic that we often want to compute is to find the size of the dataset. The dim() function outputs a vector containing the number of rows and columns or observations and variables.

```
> dim(data)
[1] 2759   11
```

Our dataset contains 2759 areas each with entries for the 11 variables listed in Table 5.1. We can alternatively use the nrow() function to find only the number of rows in the dataset.

```
> nrow(data)
[1] 2759
```

Similarly, we use the ncol() function to find the number of columns. > ncol(data) [1] 11

5.1.2 Summarizing the Data

Once we know the number of variables of the dataset, the next step in EDA is to identify the nature of the variables. The head() and tail() functions, respectively, output the first and last few entries of a data frame. This is often useful to inspect some of the values taken by the variables.

```
> head(data[,1:3])
                      NAME                          LSAD CENSUS2010POP
1         Abilene, TX Metropolitan Statistical Area        165252
2 Callahan County, TX             County or equivalent         13544
3    Jones County, TX             County or equivalent         20202
4   Taylor County, TX             County or equivalent        131506
5           Akron, OH Metropolitan Statistical Area        703200
6  Portage County, OH             County or equivalent        161419
```

The summary() function provides a brief summary of each variable.

```
> summary(data)
                                              NAME
 Carson City, NV                                :   2
 Washington-Arlington-Alexandria, DC-VA-MD-WV:   2
 Abbeville, LA                                  :   1
 Aberdeen, SD                                   :   1
 Aberdeen, WA                                   :   1
 Abilene, TX                                    :   1
 (Other)                                        :2751
                            LSAD      CENSUS2010POP        NPOPCHG_2010
 County or equivalent          :1788   Min.   :     416   Min.   :-5525.0
 Metropolitan Division         :  29   1st Qu.:   33258   1st Qu.:  -10.0
 Metropolitan Statistical Area: 366   Median :   62544   Median :   54.0
 Micropolitan Statistical Area: 576   Mean   :  239151   Mean   :  475.0
                                       3rd Qu.:  156813   3rd Qu.:  257.5
                                       Max.   :18897109   Max.   :29631.0

 NATURALINC2010     BIRTHS2010      DEATHS2010        NETMIG2010
 Min.   : -794   Min.   :    0   Min.   :    0.0   Min.   :-9786.0
 1st Qu.:     7   1st Qu.:   98   1st Qu.:   79.0   1st Qu.:  -38.0
 Median :    40   Median :  184   Median :  145.0   Median :   15.0
 Mean   :   326   Mean   :  774   Mean   :  447.9   Mean   :  148.3
 3rd Qu.:   161   3rd Qu.:  483   3rd Qu.:  318.0   3rd Qu.:  123.0
 Max.   :28129   Max.   :60868   Max.   :32739.0   Max.   :14574.0

 INTERNATIONALMIG2010 DOMESTICMIG2010      RESIDUAL2010
 Min.   :   -2.0      Min.   :-25584.00   Min.   :-204.0000
 1st Qu.:    3.0      1st Qu.:    -62.00   1st Qu.:  -3.0000
 Median :   14.0      Median :      0.00   Median :  -1.0000
 Mean   :  163.6      Mean   :    -15.34   Mean   :   0.6807
 3rd Qu.:   61.0      3rd Qu.:     73.00   3rd Qu.:   1.0000
 Max.   :20199.0      Max.   :  9416.00   Max.   : 376.0000
```

For factors or categorical variables such as NAME and LSAD, the summary contains the number of times each value occurs in the variable, ordered alphabetically. If there are too many distinct values, then the values occurring most number of times are listed at the top, and most of the values are clubbed together as (other).

One of the goals of using the summary functions is to catch errors in the data early on. The fact that Carson City, NV and Washington-Arlington-Alexandria occur twice in the dataset look suspicious. We look into the entries for these areas below.

```
> data[which(data$NAME == 'Carson City, NV'),c('NAME','LSAD')]
                   NAME                            LSAD
227 Carson City, NV Metropolitan Statistical Area
228 Carson City, NV          County or equivalent

> data[which(data$NAME == 'Washington-Arlington-Alexandria,
   DC-VA-MD-WV'),c('NAME','LSAD')]
                   NAME                            LSAD
1419 Washington-Arlington-Alexandria, DC-VA-MD-WV
    Metropolitan Statistical Area
1423 Washington-Arlington-Alexandria, DC-VA-MD-WV
    Metropolitan Division
```

The entries show that the Carson City, NV is listed twice in the dataset because it is both a metropolitan statistical area and a county. The case for Washington-Arlington-Alexandria is similar, which indicates that there is no duplication error in the dataset.

For the remaining numerical variables, the summary contains the following statistics:

1. Min.—smallest value of the variable.
2. 1st Qu. (Q1)—first quartile or 25th percentile. A quarter of the values are below this number and three quarters are above it.
3. Median—second quartile or 50th percentile. A half of the values are below this number, half are above it.
4. Mean—Average value of the variable.
5. 3rd Qu. (Q2)—third quartile or 75th percentile. Three quarters of the values are below this number and a quarter are above it.
6. Max.—largest value of the variable.

If the data has even number of elements, the median is an interpolated value between the two central elements, e.g., `median(c(1,2,3,4)) = 2.5`. A percentile of x % implies that x % of the data is below this value.

The minimum, Q1, median, Q3, and the maximum value when taken together are called as a five-number summary. These are also called order statistics, as they are specific ranks in the ordered values of the variable, for instance, using ascending order, minimum is the first position, median is the middle position, and maximum is the last position.

These statistics are useful to get a sense of the data distribution for a variable: its range and centrality. We can obtain the range of the data from the minimum and maximum values, and dispersion or how much the data is spread out between Q1 and Q3. A related statistic is the interquartile range (IQR) which is the difference between the Q3 and Q1. We can obtain the location or centrality of the data from the median and the mean.

5.1.3 *Ordering Data by a Variable*

We often want to find the extreme values of numerical variables, e.g., the top ten metropolitan areas by population. We can use the sort() function that sorts vectors or data frames by a variable. We can sort the CENSUS2010POP variable using:

```
> sort(data$CENSUS2010POP)
  [1]      416      539      690      763      783
  929      1599     1901 1987     2044     2156
  2470      2513     2790     2966     3331
  ...
```

To sort the variable in a descending order, we need to use the decreasing=T parameter.

```
> sort(data$CENSUS2010POP,decreasing=T)
  [1] 18897109 12828837 11576251  9818605
  9818605  9461105  7883147  6371773
  5965343  5946800  5582170  5564635  5268860
  ...
```

The sort function also works for string, factor, and ordered variables. The string and factor variables are sorted alphabetically, and ordered variables are sorted according to their order. The sort() function also handles missing values; by default the NA values are removed and we get a shorter sorted vector as output. To get a sorted vector with the NA values, we apply the na.last = T argument that places the NA values at the end of the vector.

In the examples above, the sort function only sorts the given variable by value. Sometimes, we need to sort a data frame by a variable. We can do this using sort(); by the passing the index.return=T argument, we obtain a data frame containing the sorted values and a vector containing indices of the sorted values. We later use this vector to index the data frame.

```
> output = sort(data$CENSUS2010POP,decreasing=T,index.return=T)
> data[output$ix[1:10],1:2]
                                                NAME
946  New York-Northern New Jersey-Long Island, NY-NJ-PA
790                 Los Angeles-Long Beach-Santa Ana, CA
962                   New York-White Plains-Wayne, NY-NJ
791                  Los Angeles-Long Beach-Glendale, CA
792                              Los Angeles County, CA
271                  Chicago-Joliet-Naperville, IL-IN-WI
272                        Chicago-Joliet-Naperville, IL
368                      Dallas-Fort Worth-Arlington, TX
1046        Philadelphia-Camden-Wilmington, PA-NJ-DE-MD
600                       Houston-Sugar Land-Baytown, TX
                                LSAD
946  Metropolitan Statistical Area
790  Metropolitan Statistical Area
962          Metropolitan Division
791          Metropolitan Division
792           County or equivalent
```

```
271  Metropolitan Statistical Area
272           Metropolitan Division
368  Metropolitan Statistical Area
1046 Metropolitan Statistical Area
600  Metropolitan Statistical Area
```

We indexed the data frame using the first 10 elements of the index vector output$ix[1:10] to obtain the names of the top-10 areas by population. The values printed to the left of the entries are the row numbers where that entry occurs in the original data frame `data`.

This method of ordering data frames is a bit convoluted; *R* provides a function called order() that orders the data frame for a given variable in one step.

```
> data[order(-data$CENSUS2010POP)[1:10],1:2]
```

This returns exactly the same output as above. The minus sign in front of the data$CENSUS2010POP variable negates it, which causes it to be sorted in reverse. The order function can also sort on multiple variables together, to do so, we can call it as `order(variable1, variable2, ...)`. This will sort the data frame by variable1, and in case there are ties, it will break them by sorting on variable2. To sort some variables in descending order, simply negate them with a minus sign: `order(variable1, -variable2, ...)`.

5.1.4 Group and Split Data by a Variable

In the examples for the sort function, we see that the output consists of all four types of areas: counties, metropolitan divisions, micropolitan statistical area, and metropolitan areas as encoded by the LSAD variable. We often need to compute statistics separately for these area types. Using the function,

```
> which(data$LSAD == 'Metropolitan Statistical Area')
```

we can first select a subset of the data related to metropolitan statistical areas, and then perform the analysis on that. That will require us to apply this function with different values of the LSAD variables. This may suffice for the our dataset, but it is cumbersome if the variable has too many values.

As we saw in the previous chapter, the by() function allows us to perform operations on subgroups of data based on values of a variable. We use by() to compute the mean population of the area types given by the LSAD variable.

```
> by(data$CENSUS2010POP,data$LSAD,mean)
data$LSAD: County or equivalent
[1] 161779.3
-------------------------------------------------------
data$LSAD: Metropolitan Division
[1] 2803270
-------------------------------------------------------
data$LSAD: Metropolitan Statistical Area
```

```
[1] 705786.2
------------------------------------------------------------
data$LSAD: Micropolitan Statistical Area
[1] 53721.44
```

There is a similar function called split() that splits the data frame according to the values of the variable. We can then compute the statistic of interest on the smaller data frames, which in our case is identifying the top five areas by population. We do this computation below by looping over the data splits.

```
> data.split = split(data,data$LSAD)
> for (x in names(data.split)) {
     dd = data.split[[x]]
     print(x)
     print(dd[order(-dd$CENSUS2010POP)[1:5],c(1,3)])
}

[1] "County or equivalent"
                       NAME CENSUS2010POP
792  Los Angeles County, CA       9818605
273         Cook County, IL       5194675
606       Harris County, TX       4092459
1062    Maricopa County, AZ       3817117
1239   San Diego County, CA       3095313

[1] "Metropolitan Division"
                                          NAME CENSUS2010POP
962             New York-White Plains-Wayne, NY-NJ      11576251
791            Los Angeles-Long Beach-Glendale, CA       9818605
272                  Chicago-Joliet-Naperville, IL       7883147
1423 Washington-Arlington-Alexandria, DC-VA-MD-WV       4377008
369                        Dallas-Plano-Irving, TX       4235751

[1] "Metropolitan Statistical Area"
                                                NAME CENSUS2010POP
946  New York-Northern New Jersey-Long Island, NY-NJ-PA      18897109
790                Los Angeles-Long Beach-Santa Ana, CA      12828837
271                Chicago-Joliet-Naperville, IL-IN-WI       9461105
368                    Dallas-Fort Worth-Arlington, TX       6371773
1046       Philadelphia-Camden-Wilmington, PA-NJ-DE-MD       5965343

[1] "Micropolitan Statistical Area"
                                   NAME CENSUS2010POP
2523                        Seaford, DE        197145
2639                     Torrington, CT        189927
2006 Hilton Head Island-Beaufort, SC        187010
2004                           Hilo, HI        185079
1781          Daphne-Fairhope-Foley, AL        182265
```

Inside the for loop, we first use split() to split the data by the LSAD variable. This function outputs a list of data frames that are stored in the variable data.split and are indexed by the value of the LSAD variable. For instance, we access the data frame containing Metropolitan Division by data.split[['Metropolitan Division']].

We iterate over all the data frames contained in the list `data.split`. To do so, we use the names() function that returns a list of data frame names.

```
> names(data.split)
[1] "County or equivalent"            "Metropolitan Division"
[3] "Metropolitan Statistical Area" "Micropolitan Statistical Area"
```

Apart from a list of data frames, the names() function is also useful to list the variables in a data frame.

In each iteration of the for loop, the x variable will contain one of the names, and we obtain the corresponding data frame by data.split[[x]]. We then use the order function to output the top five areas by population in that data frame.

We use the output of the split() function to create data frames corresponding to different area types.

```
> data.county = data.split[['County or equivalent']]
> data.division = data.split[['Metropolitan Division']]
> data.metro = data.split[['Metropolitan Statistical Area']]
> data.micro = data.split[['Micropolitan Statistical Area']]
```

5.1.5 Variable Correlation

Correlation analysis is a useful tool to measure the relationship between the different variables of the dataset. A correlation metric between two numerical variables reflects if increasing value of a variable *co-occurs* with increasing value of another variable, e.g., increase in the consumption of sweetened sodas and increase in the prevalence of diabetes [5].

In general, knowing the correlation between variables is valuable information, e.g., in a business, we can check if the advertising costs for different products correlated with their sales.

The term *co-occurrence* is of importance in the previous paragraph; we did not say that increase in the variable x *causes* an increase in variable y, i.e., consuming large amounts of high-fructose corn syrup causes diabetes. The data merely indicates that over time people have been drinking a lot of sweetened sodas and there has also been an increase in the percentage of diabetes in the population. Hence, the classic phrase "correlation does not imply causation."

Relating correlation to causation leads one swiftly to a statistical minefield. Over the last 100 years, the number of people wearing hats has certainly decreased, and the average shoe sizes have increased [4]. This does not imply not wearing hats causes big feet.

There are various versions of the correlation metric, here, we use the Pearson's correlation coefficient. A correlation metric usually takes a value between -1 and 1. A value of zero indicates no correlation between the two variables, while a larger positive or negative value indicates that the variables are correlated positively or negatively, respectively.

Means and Variances

In order to define correlation mathematically, we first formally introduce more fundamental statistics such as mean and variance.

The mean of a variable x, denoted by $\bar{x}$ or μ and computed by the *R* function mean(x), is the average value of the variable. For a variable x with n data points $\{x_1, \ldots, x_n\}$,

$$mean(x) = \frac{1}{n} \sum_{i=1}^{n} x_i.$$

The variance of the variable x, denoted by σ^2 given by the *R* function *var*(x) is a measure of how much the data is spread out from the mean. Higher the spread of the data, higher the variance.

$$var(x) = \frac{1}{n-1} \sum_i (x_i - mean(x))^2.$$

The variance is a second degree quantity, as we can see by the squared term in the summation above. We often need a metric indicating the spread of the data in the same order of magnitude as the mean. The square root of the variance is the standard deviation, denoted by σ and computed by the function *sd*(x).

$$sd(x) = \sqrt{var(x)}.$$

Covariance is the extension of variance to two variables. It is a measure of whether two variables spread out similarly. We can calculate it using the *cov*(x,y) function defined as:

$$cov(x, y) = \frac{1}{n-1} \sum_i \big(x_i - mean(x)\big)\big(y_i - mean(y)\big).$$

Correlation between two variables x and y is given by the covariance between those two variables normalized by their individual variances.

$$cor(x, y) = \frac{cov(x, y)}{\sqrt{var(x)var(y)}}.$$

We use the *cor*() function to find the covariance of a pair of variables, in this case, the census population and number of births for metropolitan areas.

```
> cor(data.metro$CENSUS2010POP,data.metro$BIRTHS2010)
[1] 0.9955464
```

We see that these two variables are highly correlated. This makes sense, as the areas with large populations should also have more number of births every year. This logic does not hold true for all variables though, e.g., we see a negative correlation

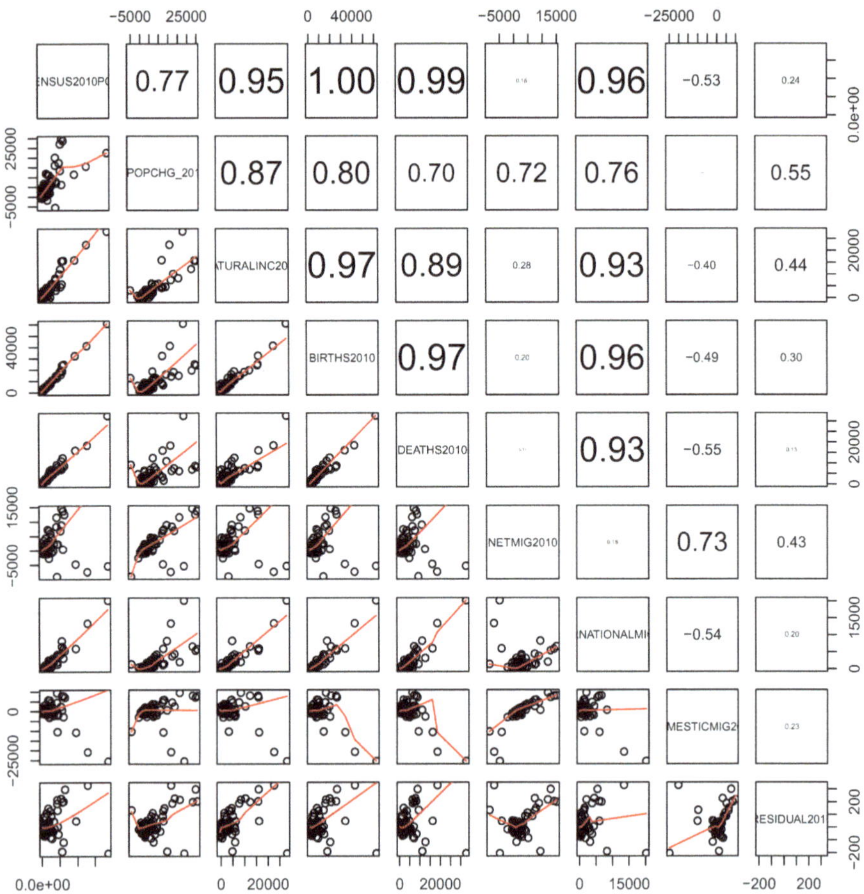

Fig. 5.1 Pairwise correlation for metropolitan statistical areas

between census population and domestic migration. This implies people are leaving areas with large population for smaller ones [3].

```
> cor(data.metro$CENSUS2010POP,data.metro$DOMESTICMIG2010)
[1] -0.5271686
```

We can use the *cor*() function to obtain the pairwise correlation between a set of numeric variables of a data frame (output omitted for conciseness).

```
> cor(data.metro[,3:11])
```

It is often better to visualize the pairwise correlation between variables: The pairs() function generates a scatter plot between all variables tiled together by default. The pairs() function is configurable; we can also specify our own functions to obtain different plots at the upper, lower, and diagonal panels.

In Fig. 5.1, we create a customized pairs plot where the lower panel contains the scatter plots with a smoothed line indicating how the variables are correlated, and the upper panel contains the correlation values. We spice up the upper panel to have the font of the entries proportional to their value.

```
# function for generating the correlation values
> panel.cor <- function(x, y, digits = 2, prefix = "", cex.cor, ...) {
    usr <- par("usr"); on.exit(par(usr))
    par(usr = c(0, 1, 0, 1))
    r <- cor(x, y)
    txt <- format(c(r, 0.123456789), digits = digits)[1]
    txt <- paste(prefix, txt, sep = "")
    if(missing(cex.cor)) cex.cor <- 1/strwidth(txt)
    text(0.5, 0.5, txt, cex = cex.cor * abs(r))
}

> pairs(data.metro[,3:11], lower.panel = panel.smooth,
        upper.panel = panel.cor)
```

5.2 Getting a Sense of Data Distribution

An important step in EDA is understanding the data distribution. Intuitively, the data distribution can be understood as a map of the data: which regions or values of the variables is the data more concentrated and which are the extreme values. In this section, we look at two powerful tools to visualize the data distribution: box plots and histograms. After that, we also look at skewness and kurtosis, which are numeric measures of data skewness.

5.2.1 Box Plots

A box plot or alternatively, box-and-whisker plot provides a visualization of the distribution of a numerical variable. It is based on the five-number summary statistics of a variable (minimum, Q1, median, Q3, maximum) that we discussed above.

Figure 5.2 shows a box plot for BIRTHS2010 variable for the entries corresponding to micropolitan statistical areas. We limit this analysis to only micropolitan statistical areas for clarity. We use the `data.micro` data frame that we created previously.

We obtain this plot using the boxplot() function.

```
> boxplot(data.micro$BIRTHS2010,names=c('BIRTHS2010'),show.names=T)
```

On the vertical axis, we have the values of the variable. *R* does not show the name of the variable on the horizontal axis by default, so we call boxplot() with the names and show.names parameters.

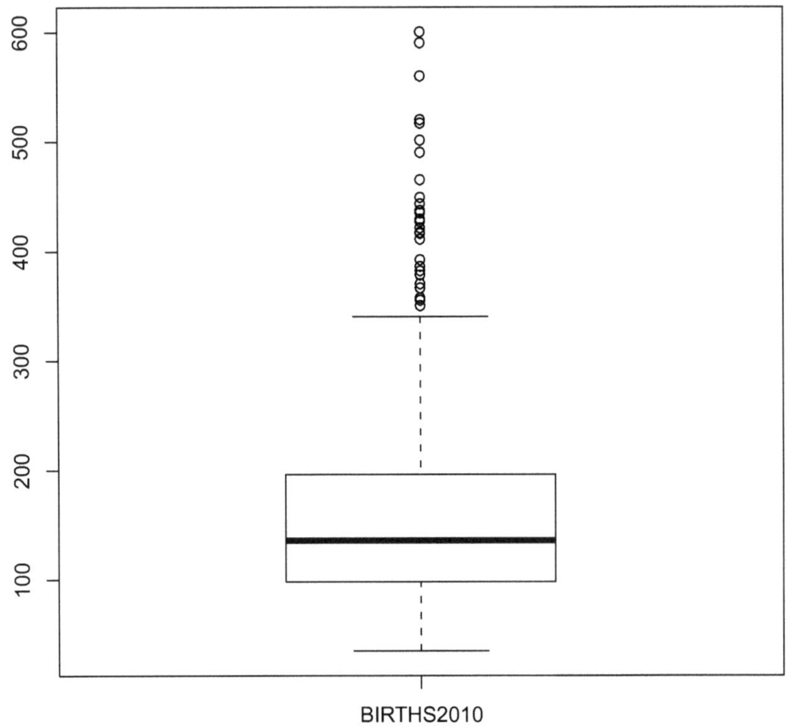

Fig. 5.2 Box plot of BIRTHS2010 for micropolitan statistical areas

A box plot has three elements:

1. Box: A rectangular box with a solid horizontal line in bold.
2. Whiskers[3]: A pair of vertical dotted lines, with solid horizontal end lines which are called notches.
3. Extremities: Unfilled circles or bubbles located above or below the whiskers.

The solid horizontal line in the box plot corresponds to the median. The lower and upper lines of the box, or whisker hinges, correspond to the 1st quartile (Q1) and 3rd quartile (Q3) respectively. The size of the box visually indicates the centrality of the data, or the proportion of the data lying close to the median.

The whisker notches correspond to the furthest data points lying in the 1.5 times the interquartile range (IQR). For a variable x, the notch of the top whisker is, therefore, min(max(x), Q3 + 1.5 * IQR), and the notch for the lower whisker is max(min(x), Q1 − 1.5 * IQR). The bubbles beyond the whiskers are the data points that are above or below the top and bottom notches, respectively. These data points are the extreme

[3] A possible explanation for the name box plot whiskers is that they resemble a cat's whiskers.

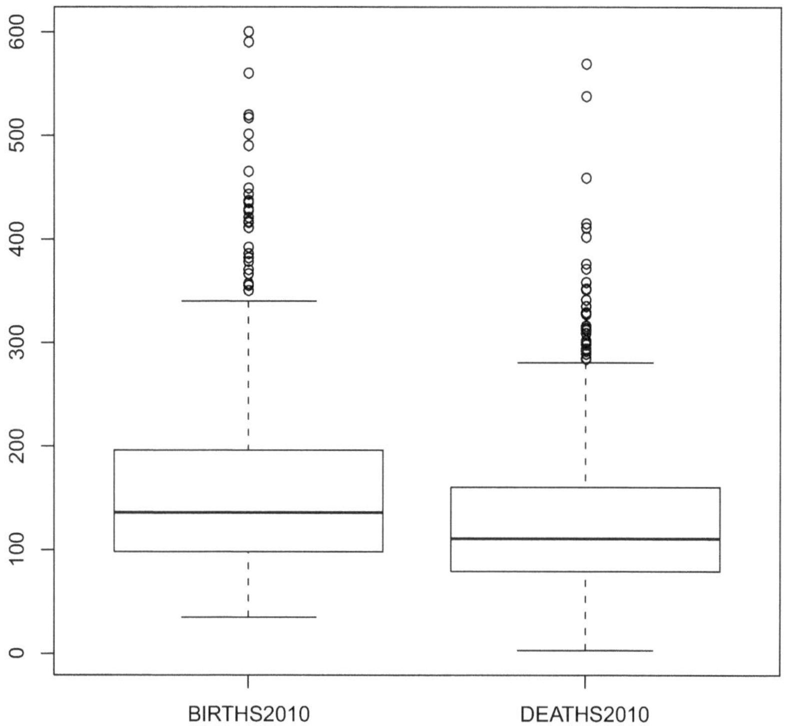

Fig. 5.3 Box plot of BIRTHS2010 and DEATHS2010 for micropolitan statistical areas

values for the variable or commonly referred to as the tail. Along with the visual representation of the data distribution, the boxplot() function also returns the numerical values of the statistics.

The box plot in Fig. 5.2 provides multiple insights into the BIRTH2010 variable. Firstly, there are extreme values only above the top whisker. This implies that dataset is well spread out, and has a few values that are much larger than the median. We can also see the same effect in the whiskers, where the top whisker is longer than the bottom whisker. Secondly, the median, denoted by the solid line, is the point at which there are equal number of data points above and below it. The short-bottom whisker indicates that a large portion of the data is concentrated in the short range between the minimum value (35) and the first quartile (98).

We can also use the boxplot() function to compare the distribution of two variables side by side. Figure 5.3 shows the box plots for BIRTHS2010 and DEATHS2010 for data.micro that we obtain using the code below.

```
> boxplot(data.micro$BIRTHS2010,data.micro$DEATHS2010,
    names=c('BIRTHS2010','DEATHS2010'))
```

We do not need to set show.names=T when we are calling box plot for multiple variables. We see that the top whisker and the extremities for BIRTHS2010 is longer

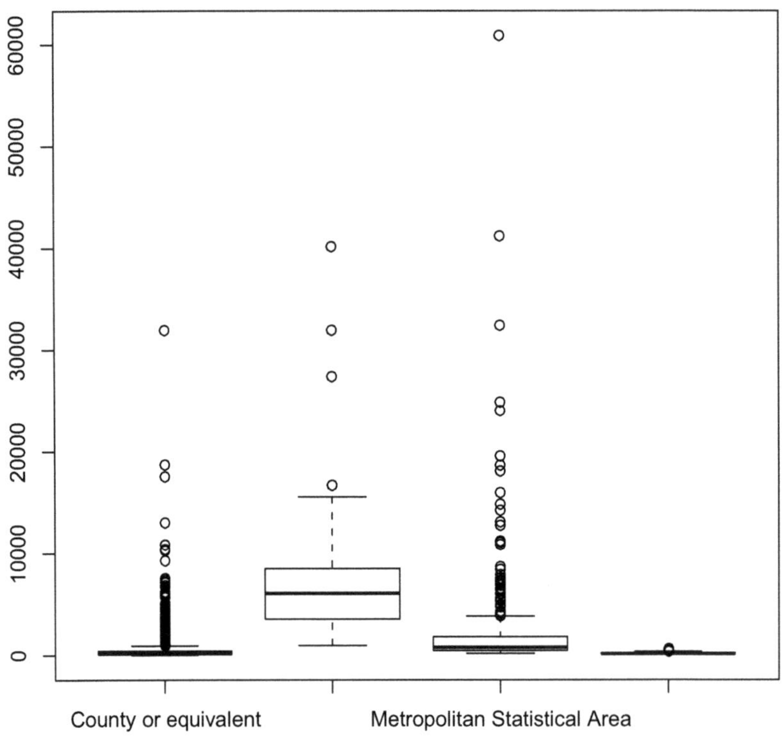

Fig. 5.4 Box plot of BIRTHS2010 for all areas

than that for DEATHS2010. This implies that the BIRTHS2010 variable is more spread out than DEATHS2010. We also see that all five statistics for BIRTHS2010 is higher than that of DEATHS2010. This implies that the the BIRTH2010 values are overall higher than DEATH2010 values.

The boxplot() function can also compare all variables in the data frame together when called with the data frame boxplot(data.micro). On the other hand, we can also compute box plots for one variable over data split across another variable. In our original data frame, the LSAD variable denotes if the entry corresponds to a county or equivalent, metropolitan area, metropolitan statistical area, or micropolitan statistical area. We compute the box plot for BIRTHS2010 over the data partition below. The output for this function is shown in Fig. 5.4.

```
> boxplot(data$BIRTHS2010 ~ data$LSAD)
```

In the figure, we have four box plots corresponding to the four area types, identified on the horizontal axis and sharing a common vertical axis. The labels for some of the areas are omitted due to lack of space: from the left, the second plot is for metropolitan divisions and the fourth is for micropolitan statistical areas. The box plots for micropolitan statistical areas appear squished; the number of births in these areas is significantly lower than those in other types of areas with a maximum value

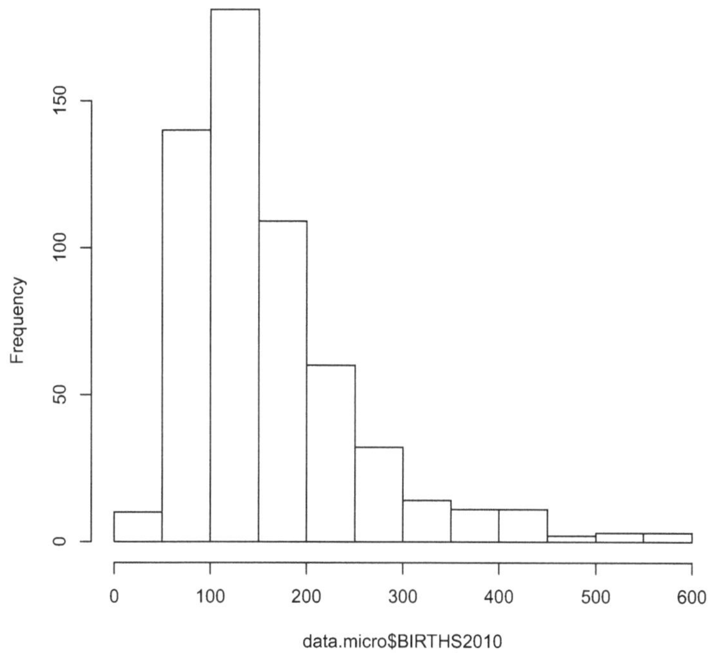

Fig. 5.5 Histogram of BIRTHS2010 for micropolitan statistical areas

of 600. It is the same box plot that we had in Fig. 5.2. The box plots for counties and metropolitan statistical areas indicate a sharper skewness of the data with large outliers compared to the median. The number of births in the New York-Northern New Jersey-Long Island metropolitan area (60,868) is about 79 times greater than the median number of births in metropolitan areas (767.5), and is in fact greater than the individual populations of 52 % of the counties in the USA.

5.2.2 *Histograms*

Along with box plot, histogram is a powerful tool to visualize the probability distribution of the data. We can get a sense of the regions in which the data is concentrated or not by its peaks and troughs, and if it is spread out in any particular direction.

We use the hist() function to plot the histogram for the CENSUS population of the micropolitan statistical areas data below (output in Fig. 5.5). As with the boxplot() function, hist() also returns the values it uses to generate the plot.

```
> hist(data.micro$BIRTHS2010)
```

The histogram consists of a set of bars corresponding to a bucket of data points. The height of a bar is the number of data points in that bucket. In this plot, the hist() function splits the data into bins or buckets of size 50: (0–50, 50–100, ..., 550–600). We see that the height of the first bar is 10, so there are 10 data points in the range 0–50.

The histogram in Fig. 5.5 visualizes the same data we had in Fig. 5.2 from a different perspective. We can infer the same data characteristics; most data points are concentrated around the median (136). Also, the bars are clearly not of uniform height and most of the shorter bars are to the right. These are the extreme values that are much larger than the median and the box plot had represented them by bubbles above the top whisker.

The choice of bin sizes is an important parameter for plotting a histogram. The hist() function supports various formulae for this purpose which can be specified by the breaks parameter. The default is the Struges' formula, which is used for the histogram in Fig. 5.5. This formula divides the data into equal-sized k bins where $k = \lceil log_2 n + 1 \rceil$ and n is the number of data points. Other strategies are the Scott and Freedman–Diaconis (FD) formulae. We can also specify a user-defined function to compute the breaks; a simple instance of this is to just specify the number of breaks.

We plot the histograms using different break formulae below (output in Fig. 5.6).

```
> hist(data.micro$BIRTHS2010,breaks='sturges',main='Struges')
> hist(data.micro$BIRTHS2010,breaks='scott',main='Scott')
> hist(data.micro$BIRTHS2010,breaks='fd',main='Freedman-Diaconis')
> hist(data.micro$BIRTHS2010,breaks=50,main='50 bins')
```

There are a few additions that we can make to the histograms to see the data from a finer perspective. We look at some of these below.

Rug

We can further highlight the data distribution by adding a rug to the histogram. A rug is simply a series of ticks on the horizontal axis corresponding to the location of the data points. The dark shaded areas are where most of the data points are located. A rug normally decorates another plot, to show it, we use the rug() function after drawing the histogram as follows (output in Fig. 5.7).

```
> hist(data.micro$BIRTHS2010,breaks='FD')
> rug(data.micro$BIRTHS2010)
```

Density

A histogram presents a coarse view of the data distribution using the number of data points lying in a bin. Sometimes, it is more informative to visualize the smooth density of the data. The density() function uses a kernel density estimator (KDE) to obtain this output. Similar to the rug, we overlay the density on top of the histogram as follows (output in Fig. 5.7).

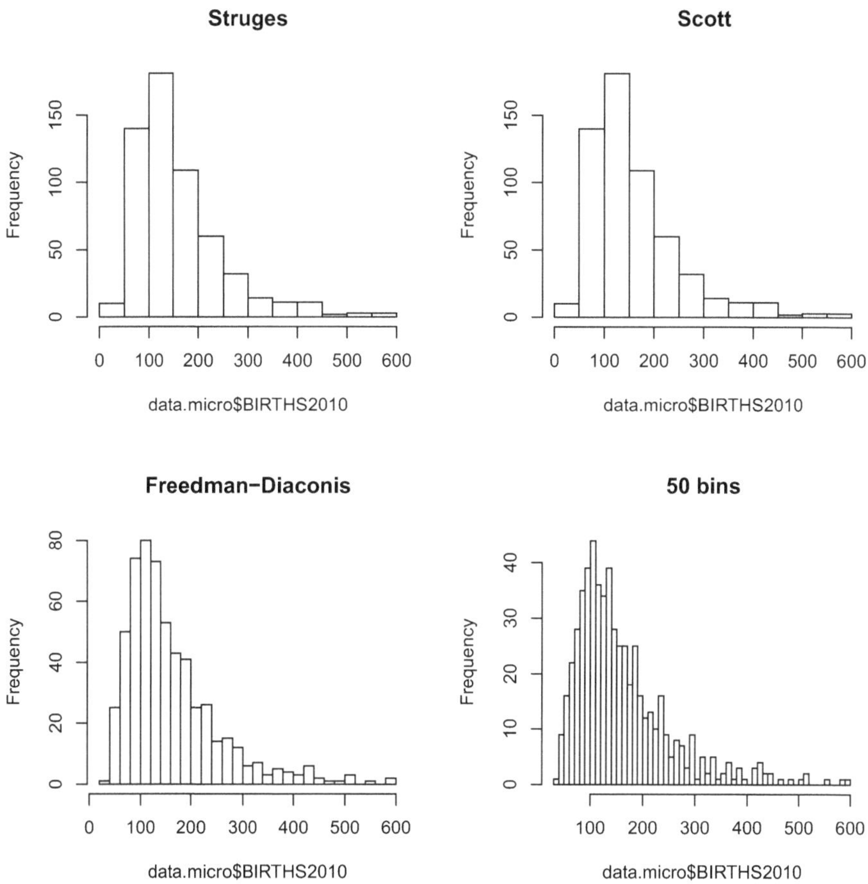

Fig. 5.6 Histogram of BIRTHS2010 for micropolitan statistical areas using different break formulae

```
> hist(data.micro$BIRTHS2010,breaks='FD',freq=F)
> points(density(data.micro$BIRTHS2010),type='l',col='red')
```

We use the freq=F flag in the hist() function to replace the vertical scale of frequencies with densities.

Cumulative Histogram

While a histogram shows the number of data points in a given bin, we often need to know the number of data points less than or equal to the range of the bin, e.g., instead of knowing how many data points lie in the bin 80–100, we need to know the number of data points lie in the range 0–100. This is called the cumulative histogram of the data, and as the name suggests, corresponds to the cumulative density function (CDF) of the data. This plot contains cumulative counts, where the count of data points in

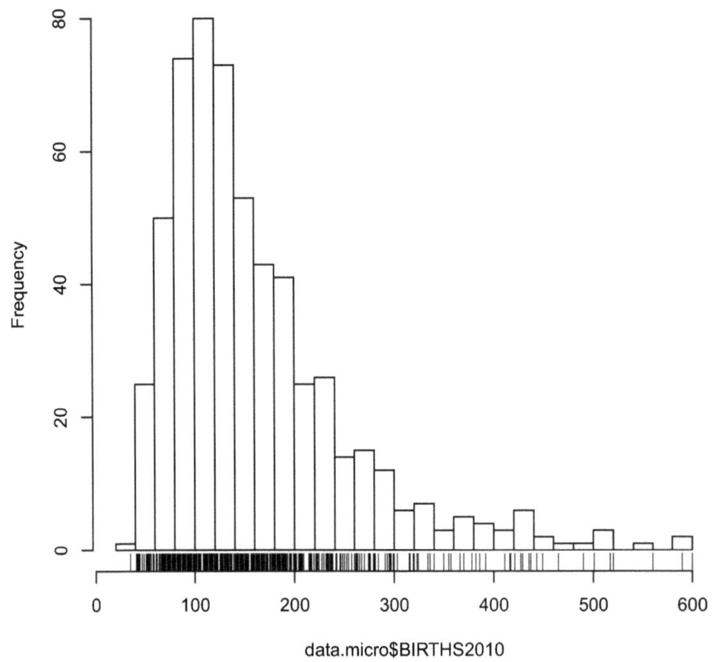

Fig. 5.7 Histogram of BIRTHS2010 for micropolitan statistical areas with a rug

bin are added to the next bin. *R* does not have a standard function to generate the cumulative histogram; we create a cumhist() function that manipulates the values returned by the histogram() function (output in Fig. 5.9).

```
> cumhist = function(x) {
    h = hist(data.micro$BIRTHS2010,'FD',plot=F)
    h$counts = cumsum(h$counts)
    plot(h)
  }
> cumhist(data.micro$BIRTHS2010)
```

We call the hist() function with the plot=F flag, because we only need the values returned by the function and not the plot. We use the cumsum() function to compute the cumulative scores and set them back to the data returned by the histogram. The cumsum() function computes a cumulative sum of a vector, e.g.,

```
> cumsum(c(1,2,3,4))
[1]  1  3  6 10
```

We finally use plot() to plot the cumulative histogram. We do not need to specify the axis labels or that we need to use bars to show the output. Being a generic function, plot() works with a histogram object and automatically displays the cumulative histogram in the right format (Fig. 5.8).

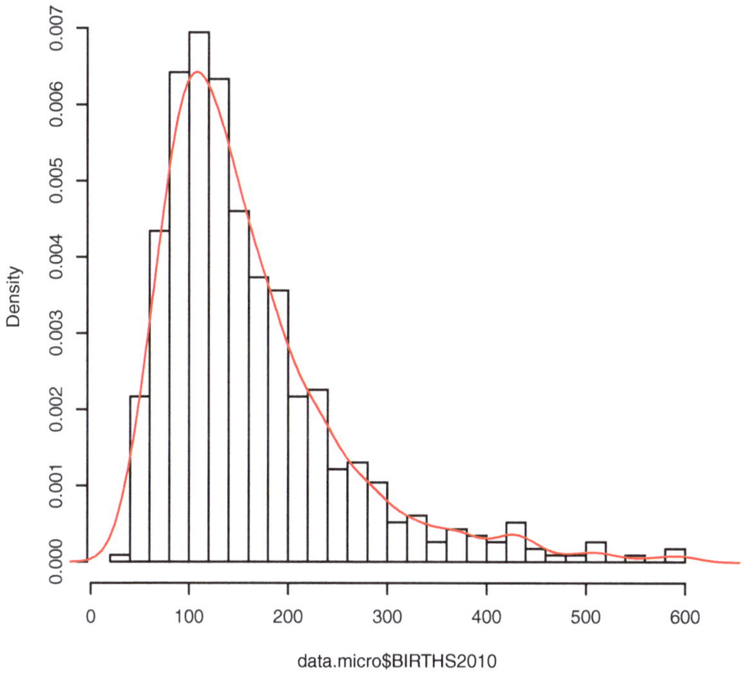

Fig. 5.8 Histogram of BIRTHS2010 for micropolitan statistical areas with smooth density function

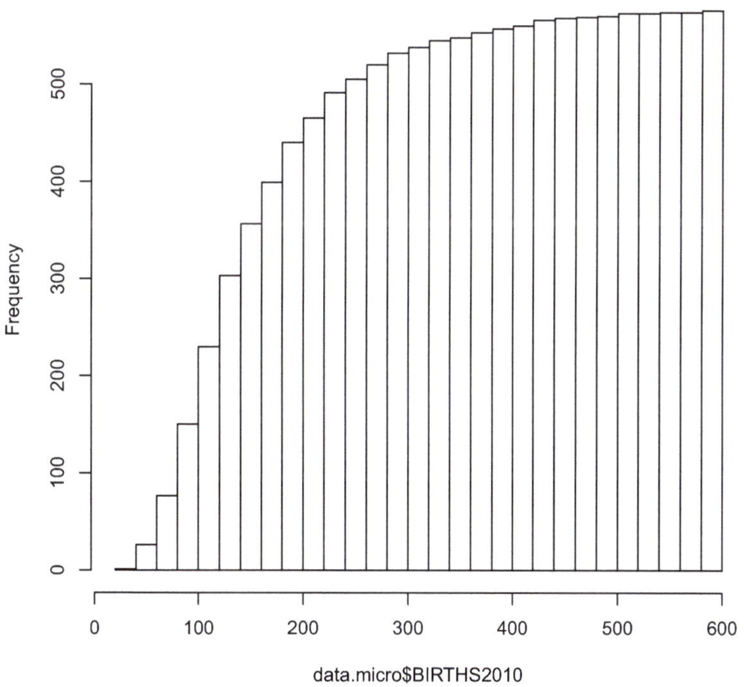

Fig. 5.9 Cumulative histogram of BIRTHS2010 for micropolitan statistical areas

5.2.3 Measuring Data Symmetry Using Skewness and Kurtosis

A histogram gives a visual representation of how the data is distributed. We often need an objective measure that summarizes the characteristic of the data. Such a measure is useful for comparing if two variables are distributed similarly.

Skewness is a metric that captures the asymmetry of a numeric variable. Mathematically, the skewness of a dataset is the third sample moment of the variable about the mean, standardized by the sample variance:

$$skewness = \frac{\frac{1}{n}\sum_i \left(x_i - \bar{x}\right)^3}{\left(\frac{1}{n}\sum_i (x_i - \bar{x})^2\right)^{\frac{3}{2}}},$$

where x_i are the individual data points and $\bar{x}$ is the mean. A variable with zero skewness implies that the data is symmetric. Positive skewness implies that the data is spread out towards the right, with extreme values larger than the median. Similarly, negative skewness implies that the data is spread out towards the left, with extreme values smaller than the median. Data sampled from power law distribution would have larger positive or negative skewness value.

We use the skewness() function in the moments package to compute the skewness of the data, for an individual variable or a data frame. We calculate the skewness of all numerical variables for micropolitan statistical areas below.

```
> library(moments)
> skewness(data.micro[,3:11])
      CENSUS2010POP          NPOPCHG_2010        NATURALINC2010
          1.7384473             2.6371220             1.0143676
         BIRTHS2010           DEATHS2010             NETMIG2010
          1.6833753             1.5502585             2.6078737
INTERNATIONALMIG2010      DOMESTICMIG2010          RESIDUAL2010
          4.4857400             2.3719011             0.9202234
```

The skewness is positive for all variables; it implies that all of them are skewed towards the right. Some variables are skewed more than others.

For comparison, we calculate the skewness of all numerical variables for metropolitan statistical areas below.

```
> skewness(data.metro[,3:11])
      CENSUS2010POP          NPOPCHG_2010        NATURALINC2010
           6.677222              4.718376              6.107227
         BIRTHS2010           DEATHS2010             NETMIG2010
           6.469855              6.606957              3.291813
INTERNATIONALMIG2010      DOMESTICMIG2010          RESIDUAL2010
           8.159912             -6.088828              3.170399
```

We see that except for DOMESTICMIG2010, the skewness is positive for all other variables. The skewness values for metropolitan areas are larger in magnitude than micropolitan areas, which implies that the data for the former is much more spread out. We see this effect in the histograms for the CENSUS2010POP variables for the two datasets in Fig. 5.10.

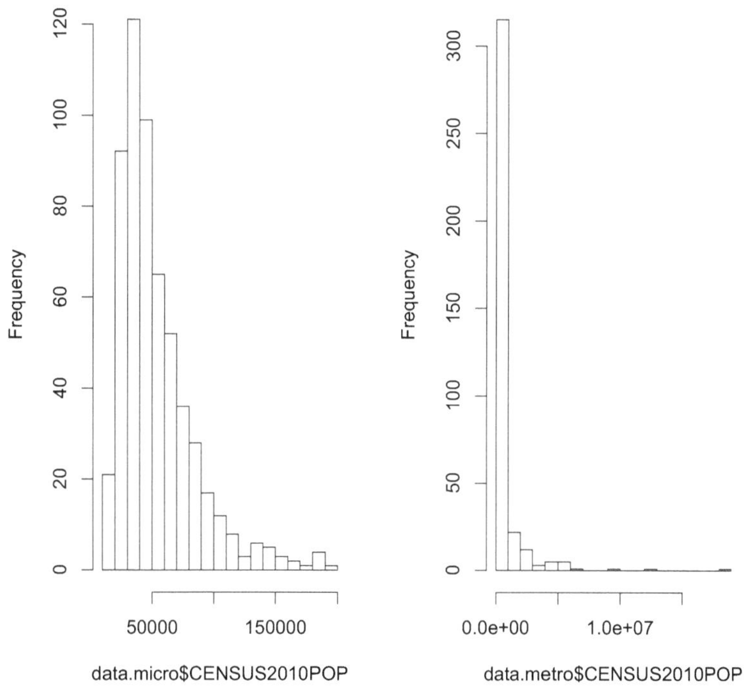

Fig. 5.10 Histograms of CENSUS2010POP for micropolitan and metropolitan statistical areas

While skewness is a measure of the degree and direction of data asymmetry, kurtosis[4] captures the sharpness or flatness of the peaks in the data: sharper the peaks, lighter the tails, and vice-versa. There are various versions of the kurtosis measure, we use Pearson's kurtosis which is the fourth sample moment of the variable about the mean, standardized by the sample variance:

$$\text{kurtosis} = \frac{\frac{1}{n}\sum_i (x_i - \bar{x})^4}{\left(\frac{1}{n}\sum_i \left(x_i - \bar{x}\right)^2\right)^2}.$$

The standard Gaussian distribution has a kurtosis of 3. This is used as a reference point; a data having lower than 3 kurtosis is more peaked than Gaussian distribution, vice versa. For that reason, the kurtosis value—3 is called excess kurtosis.[5] Uniform distribution has an excess kurtosis of 1.2.

We use the kurtosis() function in the moments package to compute the actual kurtosis of the variables for micropolitan statistical areas below.

[4] Kurtosis comes from the Greek word kurtos meaning arched.

[5] Distributions with zero, positive, and negative excess kurtosis are respectively called: mesokurtic, platykurtic, and leptokurtic distributions.

```
> kurtosis(data.micro[,3:11])
       CENSUS2010POP          NPOPCHG_2010        NATURALINC2010
            6.757994             17.459700              9.590567
          BIRTHS2010           DEATHS2010            NETMIG2010
            6.504231              6.819837             21.681844
INTERNATIONALMIG2010       DOMESTICMIG2010          RESIDUAL2010
           34.850185             22.369340             17.521871
```

There are many other measures of data symmetry, including other definitions of kurtosis. A popular measure is the Gini coefficient,[6] which captures inequality in the dataset. Gini coefficient is widely used in economics to measure the income inequality in a country. We will not go into the maths behind the Gini coefficient here.[7] We use the `gini()` function from the `reldist` package to calculate its value below.

```
> library(reldist)
> gini(data.metro$CENSUS2010POP)
[1] 0.6593656
```

5.3 Putting It All Together: Outlier Detection

Outlier detection is a good example of a data analysis task where we can apply multiple EDA techniques that we covered in this chapter. Along with missing values, outliers play a disruptive role in any nontrivial data analysis task. Outliers are the data points that are markedly deviant from rest of the data: extreme values that are either too large or too small. These values distort the actual distribution of the data, and end up having an adverse effect on our analysis.

Outliers are commonly caused due to data collection anomalies. As in the case of missing data, outliers could be caused by malfunctioning sensors or data entry errors in case of manual data collection. On the other hand, the outliers can also be genuine abnormal values occurring as a part of the data. This often indicates the presence of the data coming from a long-tailed distribution, which means there is a large probability of getting values away from the mean. Websites by their traffic, languages by their number of speakers, net-worth of individuals, are all examples of long-tailed distributions.

Also, there are cases where the identifying outliers is the goal of our analysis. This falls into the area of rare category detection, where we are interested in identifying rare, infrequently occurring data points. Examples of this include credit card fraud detection, the search for extraterrestrial intelligence (the SETI project). Simply regarding rarely occurring events as outliers sometimes leads to problems; a famous example is the detection of ozone layer over Antarctica. The software in the

[6] Named after its inventor Corrado Gini.

[7] A good reference is Dorfman [2].

NASA Nimbus 7 satellite considered low ozone values over Antarctica as outliers and automatically discarded them for many years [1].

For a given dataset, though, a question that remains is what data points can be precisely identified as outliers. There is no clear answer to that. It depends on the data distribution and the analysis methodology that we are planning to use. In the USA, a person with ten credit cards is certainly not an outlier, but someone with 1497 valid credit cards might be one.[8] An outlier is different from missing data. It does not mean that such a data point cannot exist; it merely means that including such a data point would skew our analysis. On the other hand, a person working more than 168 h a week is certainly an outlier caused due to data entry error and should be excluded from our analysis.

One approach of finding outliers is using EDA. Let us look at the survey dataset, we introduced in the previous chapter.

```
> data = read.csv('survey-fixed.csv')
> summary(data)
      sex            height          weight        handedness   exercise
 Female:118   Min.   :59.00   Min.   :  0.0   Left : 18   Freq:115
 Male  :118   1st Qu.:65.27   1st Qu.:150.0   Right:218   None: 24
 NA's  :  1   Median :67.00   Median :169.0   NA's :  1   Some: 98
              Mean   :67.85   Mean   :172.9
              3rd Qu.:70.42   3rd Qu.:191.0
              Max.   :79.00   Max.   :958.0
    smoke
 Heavy: 11
 Never:190
 Occas: 19
 Regul: 17
```

There is something anomalous about the weight variable. The maximum value is 958 lb and the minimum value of the variable is 0, which are both outliers.

In the left half of Fig. 5.11, we plot a histogram of the height variable to observe this. The histogram shows the two outliers: the lone extreme value to the far right of the majority of the data. It also shows the zero weight value to the left. We remove these values from our analysis.

The box plot is also a useful outlier detection tool. As we saw earlier, it shows the extreme values that are distant from the median as bubbles. In the right half of Fig. 5.11, we plot the box plot for the weight variable. It also shows the two outliers by the top and bottom bubbles. However, not all extreme values are outliers; while we see that a few are concentrated near the top and bottom whisker notches, these are still close to the remainder of the data as we see in the histogram.

[8] Walter Cavanagh, of Shell Beach, CA has the world record for the most number of credit cards [6].

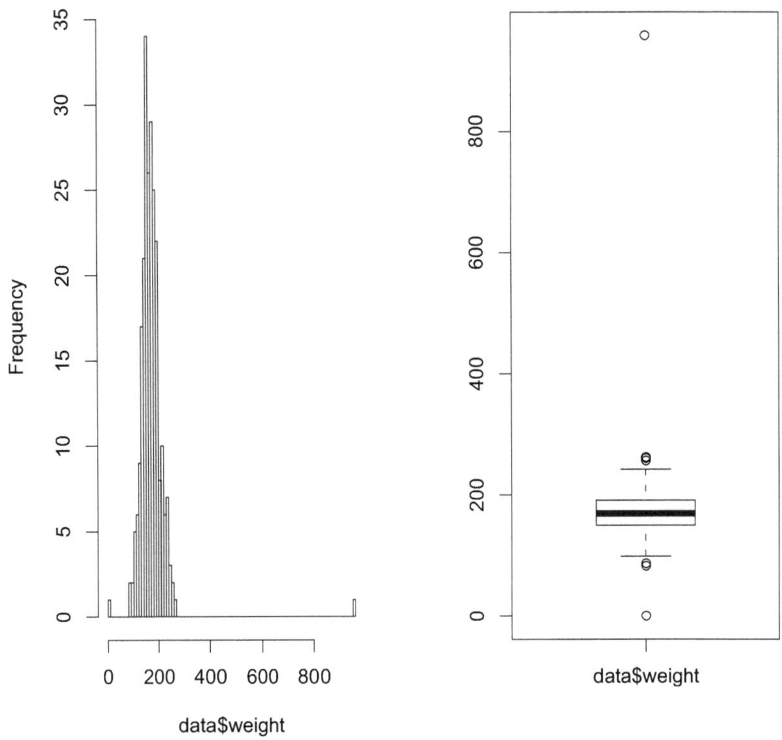

Fig. 5.11 Box plot and histogram for the weight variable

5.4 Chapter Summary

In this chapter, we looked at different techniques for EDA. Computing the summary statistics is usually the first step in EDA. This includes computing the five-number summaries of each variable: minimum, first quartile, median, third quartile, maximum along with means and standard deviations. We also looked at using the by() and split() functions to aggregate and split a data frame by a variable. To get a deeper understanding of the dataset, it is necessary to understand the underlying data distribution. Box plot is a useful method of visualizing the five-number summary of a variable along with the outliers. Histogram is a powerful tool to look deeper into the data distribution by creating a bar chart of frequencies of occurrence for different buckets of data points. We also looked at skewness() and kurtosis() as two numerical measures of the asymmetry and peakedness of the data distribution. Finally, we also look at the problem of using EDA techniques to find the outliers in the data.

References

1. Christie, M. (2001). *The Ozone layer: A philosophy of science perspective*. United Kingdom: Cambridge University Press. (Chap. 6).
2. Dorfman, R. (1979). A formula for the Gini coefficient. The review of economics and statistics. *The Review of Economics and Statistics, 61,* 146–149.
3. Most major U.S. cities show population declines. *USA Today*, June 2011.
4. Size 8 is the new 7: Why our feet are getting bigger. *Time Magazine*, Oct 2012.
5. Sugary sodas high in diabetes-linked compound. http://abcnews.go.com/Health/Healthday/story?id=4508420&page=1#.UUzdKFt34eF. March 2007.
6. To his credit, charge card king doesn't cash in. *Los Angeles Times*, Dec 2004.

Chapter 6
Regression

6.1 Introduction

In this chapter we look at the central methodology of data science: creating a statistical model. A model is a compact representation of the data and therefore of the underlying processes that produce the data. Just as architects create models of buildings to get a better understanding of how the various components of a building are laid out, data scientists create statistical models to obtain an analogous understanding of the data.

There are two types of statistical modeling tasks: descriptive and predictive. In descriptive modeling, the goal is to identify the relationships between variables to learn more about the structure of the data. In predictive modeling, the goal is to predict the value of a variable given the remainder of the variables. The variable to be predicted is called the dependent or target variable, and the variables used for prediction are called independent or predictor variables. The descriptive and predictive modeling tasks often go hand in hand: we can use the same modeling technique for both types of analysis.

All models are not equivalent though; being a compact representation, a model is almost always a simplified representation of the data. For every model, there is an underlying modeling assumption, e.g., in case of linear models, we assume a linear relationship between variables. These assumptions place limitations on the model in terms of the cases where we can apply it and how to interpret its results. It is important to be aware of the modeling assumptions, much like knowing the side-effect warnings on a medical drug label.

In this chapter, we will look at one the most commonly used statistical modeling approaches: regression. In the following chapter, we will look at classification. In regression, the target variable is numeric, e.g., inches of rain for the day, whereas in classification, the target variable is categorical, e.g., weather forecast for the day (sunny, cloudy, rain, snow). There is a whole laundry list of modeling approaches for regression; we will look at some of the fundamental and frequently used approaches.

Model building is an iterative process as there are typically multiple modeling approaches that are applicable for a given tasks. Even for a single modeling approach, there is a list of parameters which we can set to obtain different configuration of the

M. A. Pathak, *Beginning Data Science with R*,
DOI 10.1007/978-3-319-12066-9_6

model. Model validation is the task of determining how well a model is fitting our data. We use a score that we obtain from our validation methodology to compare the fit of different models.

6.1.1 Regression Models

Regression analysis involves identifying relationships between variables. Given a set of independent or predictor variables, our goal is to predict a dependent or target variable. In regression analysis, the target variable is numeric, and the predictor variables can be numeric, categorical, or ordinal.

Regression ultimately involves learning a function that computes the predicted value of the target variable given the predictor variables. Conventionally, we use the symbols $x_1, \dots, x_m$ to denote the m-predictor variables and y to denote the target variable. Assuming that the dataset contains n data points, these variables would be n dimensional vectors: $y = (y_1, \dots, y_n)$. Using this notation, we have the following functional form:

$$y = f(x_1, \dots, x_m; w),$$

where w is the set of parameters of the model. By substituting the values of $x_1, \dots, x_m$ and w in the function $f()$, we compute the value of y.

Most regression techniques, and most modeling techniques in general, involve a two step process:

1. **Model Fitting:** For a given dataset with variables $x_1, \dots, x_m$ and y we calculate the model parameters w that best meet certain criteria.

2. **Model Prediction:** Given a dataset with the predictor variables $x_1, \dots, x_m$ and model parameters w, we calculate the value of the target variable y. The dataset could be the same one we use for model fitting or it could also be a new dataset.

Depending on the choice of the $f()$ function, there are hundreds of regression techniques, each with a method for model fitting and model prediction.

The most commonly used family of techniques is linear regression, where we assume $f()$ to be a linear function. In this section we first look at some of the techniques for linear regression, and then move on to methods for nonparametric regression.

One advantage of using R is that most functions for regression have similar interfaces, so that we can switch back and forth between techniques without too much effort. This is because most regression functions have similar interfaces in terms of their inputs.

Table 6.1 List of variables in automobiles dataset

Variable name	Description
make	Make of the car
fuel.type	Car fuel type (diesel, gas)
num.doors	Number of doors
body.style	Car body style
engine.location	Front or rear
wheel.base	Length of wheel base
length	Length of the car
width	Width of the car
height	Height of the car
num.cylinders	Number of cylinders
engine.size	Size of car engine
fuel.system	Car fuel system
horsepower	Power rating of the engine
city.mpg	City mileage
price	Price of the car

Case Study: Car Prices

We illustrate various regression techniques using a dataset of cars.[1] The dataset lists various attributes of a car such as make, engine type, mileage, along with its price. Table 6.1 contains the list of variables used in this dataset. We use regression analysis to identify the relationship between the attributes of the car and its price. If we are able to predict the prices accurately, we can build our own car price estimation service like the Kelly Blue Book. For used cars, we only need add more variables such as how old is the car, number of miles, and the number of owners.

We load and attach the dataset in the code below.

```
> data = read.csv('autos.csv')
> attach(data)
```

6.2 Parametric Regression Models

Statistical models are of two types: parametric, where we assume a functional relationship between the predictor and target variables, and nonparametric, where we do not assume any such relationship. We will first look at techniques for parametric regression.

[1] Data obtained from 1985 Ward's Automotive Yearbook and is also located in the UCI Machine Learning repository. http://archive.ics.uci.edu/ml/datasets/Automobile

6.2.1 Simple Linear Regression

Simple linear regression is a special case of linear regression with only one predictor variable x and one target variable y. In a dataset with n data points, these variables are n-dimensional vectors $y = (y_1, \dots, y_n)$ and $x = (x_1, \dots, x_n)$. In the next subsection we will look at linear regression with multiple predictor variables. In linear regression, we assume a linear relationship between target and predictor variables.

$$y = f(x; w) = w_0 + w_1 x + \varepsilon. \tag{6.1}$$

The regression model has two *coefficients*: $w = (w_0, w_1)$, and the *error term* ε. We also refer to the coefficients as the *intercept* and the *slope*, respectively. The linear relationship between the variables is the underlying assumption of the linear regression model. This relationship is given by the slope parameter; if the value of the x variable changes by 1 unit, the predicted value of the y variable will change by w_1 units. This is not the case for a nonlinear model like $y = w_1/x$ or $y = \sin(w_1 x)$.

As we discussed earlier, a statistical model is a simplification of the data. A simple linear regression model represents our dataset of two n-dimensional variables x and y by two numbers w_0 and w_1. The error term ε is the error in our prediction. For each data point pair (x_i, y_i), we have a corresponding error or *residual* ε_i given by rearranging Eq. 6.1.

$$\varepsilon_i = y_i - w_0 - w_1 x_i. \tag{6.2}$$

We desire a model that has the minimum error. For that purpose, in the model fitting step, we need to choose the model coefficients having the minimum error. In the linear regression, we obtain the coefficients having the minimum squared error which is also called the least-squares method. The function that we are trying to minimize, in this case the sum of squared residuals, is called the *loss function*.

$$w^* = \arg\min_{w} \sum_i \varepsilon_i^2 = \arg\min_{w_0, w_1} \sum_i (y_i - w_0 - w_1 x_i)^2.$$

We obtain the following expressions for the optimal values of w_0 and w_1 using differential calculus.[2]

$$\begin{aligned} w_1^* &= \frac{\sum_i (x_i - \bar{x})(y_i - \bar{y})}{\sum_i (x_i - \bar{x})^2}, \\ w_0^* &= \bar{y} - w_1^* \bar{x}, \end{aligned} \tag{6.3}$$

where $\bar{x}$ and $\bar{y}$ are the means of variables x and y, respectively.

---

[2] We set the derivative of the loss function to zero, and solve for w_0 and w_1.

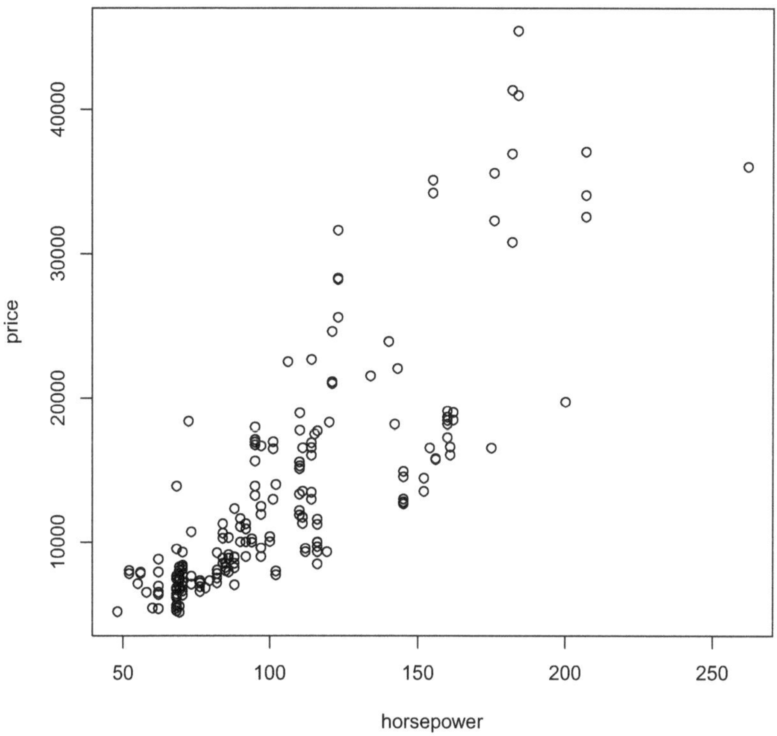

Fig. 6.1 A scatter plot between horsepower and price

6.2.1.1 Model Fitting Using lm()

In R we do not need to write code to compute the linear regression coefficients ourselves. As with most statistical functions, R has built-in support for linear regression through the `lm()` function.[3] This function takes a formula that denotes the target and the predictor variables and outputs a linear regression model object.

With this background, we perform the simple linear regression analysis on the automobiles dataset. We consider `horsepower` as a predictor variable to predict the car price. Intuitively, a more powerful car should have a higher price. We visualize this property in the scatter plot shown in Fig. 6.1.

In the code below, we use the `lm()` function to perform linear regression with the formula `price ~ horsepower`. We store the model returned by the `lm()` function in the `model` object.[4]

[3] lm() stands for linear model.

[4] By default printing the model object displays the rounded value of coefficients. For precision, it is advisable to use the coefficient values obtained from the coef() function.

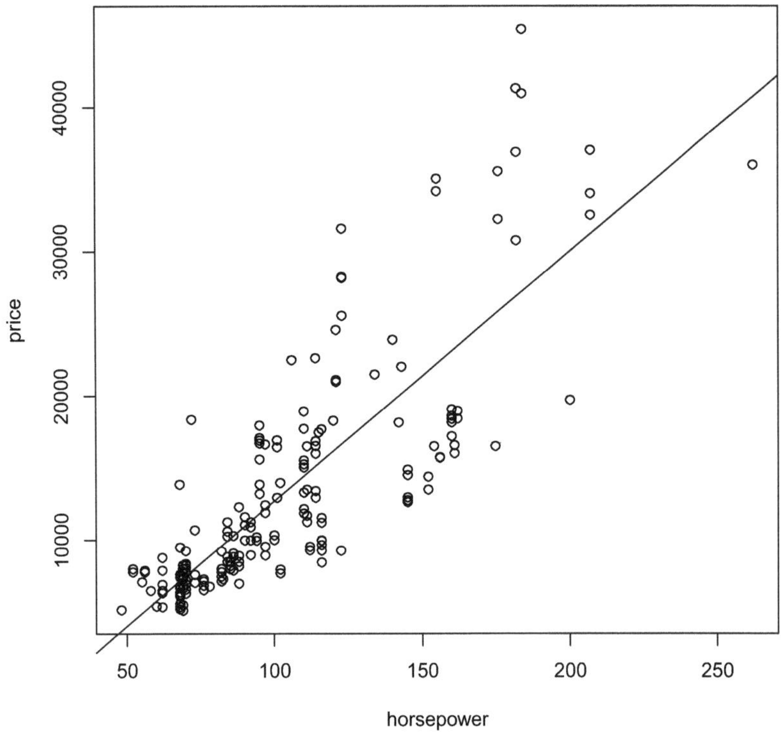

Fig. 6.2 A scatter plot between horsepower and price with the linear regression line

```
> model = lm(price ~ horsepower)
> coef(model)
(Intercept)  horsepower
 -4630.7022    173.1292
```

We observe the model coefficients by printing the model object. Here, we have the intercept $w_0 = -4630.7022$ and the slope $w_1 = 173.1292$. The model indicates the following relationship between the horsepower rating of the car and its predicted price.

$$price = 173.1292 \times horsepower - 4630.7022 \tag{6.4}$$

We can visualize this linear relationship between `horsepower` and `price` as a line on the scatter plot between these two variables. The slope and intercept of this line is the model slope and intercept. We do not need to draw a line with these parameters manually; we call the `abline()` function with the `model` object after drawing the scatter plot. Fig. 6.2 shows the output.

```
> plot(horsepower,price)
> abline(model)
```

We make two observations here: firstly, as the slope is positive, the predicted price of a car increases with its horsepower rating by a factor of \$ 173.13 per horsepower. This is consistent with our intuition that more powerful cars have higher prices.

The second observation is about the negative intercept term in the model. We might naively infer that the predicted price of a car with zero horsepower is \$ -4630.70. This does not make sense because in the real world, no car will ever have a negative price, where the seller pays the buyer an amount to buy the car. The reason for this inconsistency is that we never see cars with near-zero horsepower in our dataset, or for that matter in car dealerships [1]. The least powerful car that we have in our dataset is a Chevrolet with 48 hp. For that car, our predicted price would be \$ 3679.50, which is not unrealistic.

It is not necessary to have an intercept term in the linear regression model. The interpretation of a model without an intercept is that when the predictor variable is zero, the predicted value is also zero. Geometrically, this is equivalent to having the regression line passing through the origin. R formulae have special notation for models without the intercept: we specify `- 1` as a predictor variable. We obtain a model without the intercept to predict price given horsepower using the formula `price ~ horsepower - 1`.

```
> lm(price ~ horsepower - 1)

Call:
lm(formula = price ~ horsepower - 1)

Coefficients:
horsepower
     133.7
```

The slope of this model is 133.7, which implies that we have the following relationship between horsepower and price.

$$\text{price} = 133.7 \times \text{horsepower} \tag{6.5}$$

It is interesting to compare the models with and without the intercept. The model with the intercept has a larger slope because of the large negative intercept term: -4630.7022. The latter model has a smaller slope because of the absence of this term. Smaller slope implies a flatter line; the predicted prices for lower horsepowers will be higher than those for higher horsepowers (Fig. 6.3).

6.2.1.2 Modeling Categorical Variables

In linear regression, we can also have categorical variables as predictors. We perform simple linear regression on the car price and `make` below. The `make` variable takes 21 values as shown below.

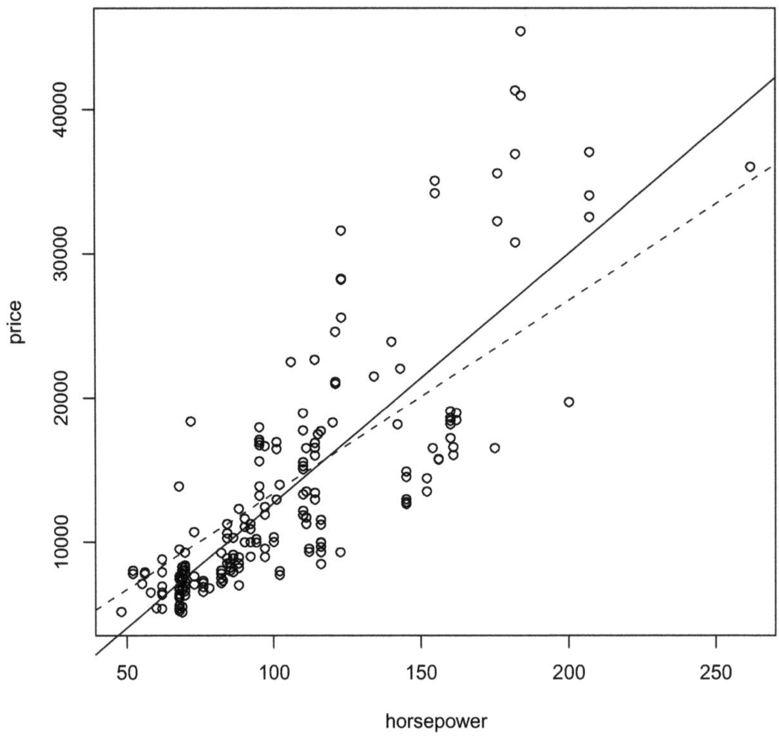

Fig. 6.3 A scatter plot between horsepower and price and linear regression lines for the two models: with intercept (*solid*), without intercept (*dashed*)

```
> unique(make)
 [1] alfa-romero  audi         bmw     chevrolet   dodge
 [6] honda        isuzu        jaguar  mazda       mercedes-benz
[11] mercury      mitsubishi   nissan  peugot      plymouth
[16] porsche      saab         subaru  toyota      volkswagen
[21] volvo
21 Levels: alfa-romero audi bmw chevrolet dodge honda isuzu
   jaguar ... volvo
```

There is no difference in syntax when using numeric and categorical variables in the formula. We use `lm()` to perform the regression using the formula `price ~ make`.

```
> model.cat = lm(price ~ make)

Call:
lm(formula = price ~ make)

Coefficients:
(Intercept)              makeaudi          makebmw    makechevrolet
 15498.333               2360.833        10620.417        -9491.333
```

```
makedodge           makehonda        makeisuzu       makejaguar
-7708.208           -7313.641        -6581.833        19101.667
makemazda   makemercedes-benz     makemercury   makemitsubishi
-5646.333           18148.667         1004.667        -6258.564
makenissan           makepeugot   makeplymouth      makeporsche
-5082.667              -9.242        -7534.905        15902.167
makesaab          makesubaru      maketoyota    makevolkswagen
-275.000           -6957.083        -5612.521        -5420.833
makevolvo
2564.848
```

Instead of assigning a coefficient to the `make` variable, the model assigns coefficients to the individual values of the variable, except for the first one. Internally, the `lm()` function represents the `make` variable as 20 binary variables; the price for a car will be the coefficient for the make of that car plus the intercept. We assume the coefficient of the make Alfa-Romero to be zero, so the price of a Alfa-Romero car would be equal to the intercept.

We can see the underlying matrix of variables that are used by `lm()` with the `model.matrix()` function. The matrix is a data frame with 21 columns, including the intercept which has always value 1, and the remaining 20 columns corresponding to the values of the make variable. The rows correspond to the cars in the original data frame and the entry corresponding to the make of the car is set to 1.

```
> model.matrix(model.cat)
    (Intercept) makeaudi makebmw makechevrolet makedodge ...
1             1        0       0             0         0
2             1        0       0             0         0
3             1        0       0             0         0
4             1        1       0             0         0
5             1        1       0             0         0
...
```

The model matrix for numeric variables, e.g., `horsepower` will only contain the variable and the intercept.

```
> model.matrix(lm(price ~ horsepower))
    (Intercept) horsepower
1             1        111
2             1        111
3             1        154
4             1        102
5             1        115
...
```

We can also create a linear regression model for a categorical variable without an intercept. In this case, the coefficients are assigned to all values of the variable.

```
> lm(price ~ make - 1)

Call:
lm(formula = price ~ make - 1)
```

```
Coefficients:
makealfa-romero     makeaudi           makebmw       makechevrolet
15498               17859              26119           6007
makedodge           makehonda          makeisuzu       makejaguar
7790                8185               8916            34600
makemazda  makemercedes-benz           makemercury   makemitsubishi
9852               33647               16503           9240
makenissan          makepeugot         makeplymouth  makeporsche
10416               15489              7963            31400
makesaab          makesubaru           maketoyota    makevolkswagen
15223                8541              9886            10078
makevolvo
18063
```

The predicted price of a car will be equal to the model coefficient. It is interesting to note that for a categorical variable, simple linear regression has an effect of averaging. The model coefficient for a value of `make` is the average price of cars of that `make`.

```
> by(price, make, mean)
make: alfa-romero
[1] 15498.33
------------------------------------------------------------
make: audi
[1] 17859.17
------------------------------------------------------------
make: bmw
[1] 26118.75
------------------------------------------------------------
make: chevrolet
[1] 6007
------------------------------------------------------------
make: dodge
[1] 7790.125
------------------------------------------------------------
make: honda
[1] 8184.692
...
```

6.2.1.3 Model Diagnostics

Once we obtain a model, the next question is to know how well it fits the data. The model diagnostics are useful in two respects: firstly, we obtain a quantitative score for the model fit, which we can use to compare different models fitted over the same dataset to see which works best. Secondly, at a fine level, we can also obtain more insight into the contributions of individual variables to the model, beyond their coefficients. This is useful in selecting which variables to use in a model.

Apart from the coefficients, the `model` object stores additional information about the linear regression model such as the residual statistics and standard error. We use the `summary()` function to look under the hood of a model. The `summary()` function is *generic*; as we have seen in Chap. 5 this function also provides summary statistics for data frames and variables. Apart from linear regression, this function also provides a summary of other types of models.

We obtain the summary of the simple linear regression model between price and horsepower that we fitted above.

```
> summary(model)

Call:
lm(formula = price ~ horsepower)

Residuals:
     Min        1Q    Median        3Q       Max
-10296.1   -2243.5    -450.1    1794.7   18174.9

Coefficients:
            Estimate Std. Error t value Pr(>|t|)
(Intercept) -4630.70     990.58  -4.675 5.55e-06 ***
horsepower    173.13       8.99  19.259  < 2e-16 ***
---
Signif. codes:  0 *** 0.001 ** 0.01 * 0.05 . 0.1   1

Residual standard error: 4728 on 191 degrees of freedom
Multiple R-squared: 0.6601, Adjusted R-squared: 0.6583
F-statistic: 370.9 on 1 and 191 DF,  p-value: < 2.2e-16
```

The `summary()` function prints the five-number summary of the model residuals, which are the differences between the values of the target variable and the values predicted by the model. These are the errors that the model makes in prediction. We can also obtain the residuals for a model using the `residuals()` function.

```
> residuals(model)
          1           2            3          4           5
 -1091.63423  1913.36577  -5531.18797  921.52818   2170.84914
...
```

We can also obtain a visual interpretation of the residuals by plotting the model object.

The residual standard error is a measure of how well the model fits the data or *goodness of fit*. It is the square root of the sum or squared residuals normalized by the degrees of freedom, which is the difference between the number of data points and variables. It is closely related to the root mean squared error (RMSE), which is another statistic used to measure the model fit error. For the above model, the residual standard error is 4728. We would like to have this as low as possible.

The R-squared (R^2) statistic is another measure of model fit. R^2 is also called the coefficient of determination: it indicates how well the predictor variable is able to capture the variation in the target variable about its mean. R^2 takes values on a scale of 0–1; a score of 0 indicates that the predictor variable does not determine the target variable at all, whereas 1 indicates that the predictor variable perfectly determines the target variable. The above model has an R^2 of 0.6601, indicating that horsepower is able to explain over 66 % of the variance in price. Adjusted R^2 statistic is the R^2 statistic weighted by a factor involving the degrees of freedom.

The model summary also provides more information about the predictor variables. Apart from the model coefficient for a variable, we also obtain its standard error and a measure of confidence given by a t value. Interpreting these terms requires a deeper understanding of how linear regression works. Model fitting involves estimating the distribution of the model coefficients. The *t value* is a function of the model coefficient and the standard error.[5] The $Pr(> |t|)$ term indicates the *p value* which is the probability of the model coefficient not being zero. This probability term provides a measure of confidence of about the importance of the variable; if the model coefficient is zero, this predictor variable is absent from our model.

6.2.1.4 Model Prediction Using Predict()

We use the `predict()` function to obtain model predictions as given by Eq. 6.4. The `predict()` function is *generic*; apart from linear regression, it provides predictions for most types of models. By default, the predictions are computed over the dataset we used to fit the model.

```
> predict(model)
        1         2         3         4         5         6
14586.634 14586.634 22031.188 13028.472 15279.151 14413.505
...
```

We can also call `predict()` with another data frame to obtain its predictions. The only requirement is that the new data frame should also have the same predictor variable, which in this case is `horsepower`.

```
> new.data = data.frame(horsepower = c(100,125,150,175,200))
> predict(model,new.data)
       1        2        3        4        5
12682.21 17010.44 21338.67 25666.90 29995.13
```

Just as we have a measure of confidence for the model coefficients, we can also obtain confidence intervals (CIs) on the predicted values. The CI as the range in which the *mean* predicted value lies with a certain probability. Conventionally, 95 and 99 % confidence levels are used in practice. This is a powerful concept; apart from having the point estimates for predicted values, the regression model gives us a probability distribution of the estimates. The distribution of the estimate has the mean equal to point estimate and variance proportional to the size of the CI. Knowing the distribution about the predictions is useful in multiple applications from actuarial studies to simulations.

We obtain CIs of predictions by calling `predict()` with the `interval='confidence'` parameter. The confidence level is controlled by the `level` parameter, which is set to 0.95 or 95 % confidence level by default.

---

[5] The name t value comes from Student's t distribution.

```
> predict(model,new.data,interval='confidence')
       fit      lwr      upr
1 12682.21 12008.03 13356.40
2 17010.44 16238.24 17782.65
3 21338.67 20275.14 22402.20
4 25666.90 24232.01 27101.79
5 29995.13 28156.72 31833.53
```

The `predict()` function returns a data frame with `fit`, `lwr`, and `upr` columns. The fit column contains the same point estimate that we obtained previously, and the `lwr` and `upr` variables contain the lower and upper limits of the CIs. For example, for the first predicted value, our estimate is 12,682.21 and the 95 % confidence interval is `[12008.03, 13356.40]`.

6.2.2 Multivariate Linear Regression

In real world applications, there are usually more than one variables that affect the target variable. Multivariate linear regression is extension of simple linear regression to the case of multiple input variables. Performing linear regression with multiple predictor variables has a different interpretation from performing multiple simple linear regression analyses with those variables. Here, we are treating the target variable as a function of the different predictor variables. By doing so, we can observe the effect of changing one predictor variable while controlling the values of the other predictor variables.

Given m input variables $x_1, \ldots, x_m$, we use the target variable y using the following linear relationship that will serve as our model.

$$y = w_0 + w_1 x_1 + \cdots + w_m x_m = w_0 + \sum_{i=1}^{m} w_i x_i. \tag{6.6}$$

Here, we have $m + 1$ model coefficients $w_0, \ldots, w_m$. Similar to simple linear regression, we obtain the coefficients for weights using differential calculus.

We also use the `lm()` function to perform multivariate linear regression by specifying the multiple predictor variables. Similarly, we use the predict() function to obtain predictions as well. In the following example, we perform linear regression on price of the car given its length, engine size, horsepower, and mileage (city MPG).

```
> lm(price ~ length + engine.size + horsepower + city.mpg)

Call:
lm(formula=price ~ length+engine.size+horsepower+city.mpg)

Coefficients:
(Intercept)      length  engine.size    horsepower      city.mpg
  -28480.00      114.58       115.32         52.74         61.51
```

This gives us a linear relationship between these five variables:

$$\begin{aligned} \text{price} =& 114.58 \times \text{length} + 115.32 \times \text{engine.size} + 52.74 \times \text{horsepower} \\ &+ 61.51 \times \text{city.mpg} - 28{,}480. \end{aligned} \tag{6.7}$$

All of the model coefficients are positive, which implies that the predicted car price increases by increasing any of these variables. This follows from our previous intuition that cars with more horsepowers and larger engines have higher prices. As length also has a positive coefficient, we observe that bigger cars are also more expensive.

We can observe the controlling aspect of multivariate regression in this example. In the simple regression analysis that we had in Eq. 6.4, the price of the car increased by a factor of \$ 173.13 per horsepower and this has gone down to a much smaller factor of \$ 52.74 per horsepower in this analysis. This is because in the simple regression analysis, we were treating price as a function of horsepower alone. In practice, the cars with more horsepower typically also have larger engines that enable them to generate that kind of power. In the multivariate regression analysis above, when considering the relationship between price and horsepower, we are controlling for length, engine size, and city MPG. We can interpret this analysis as the price of the car increases by a factor of \$ 52.74 per horsepower while keeping the engine size, car length, and mileage constant.

Mileage plays an interesting role here: if we perform simple linear regression on price vs. city MPG, we observe that city MPG has a negative coefficient.

```
> lm(price ~ city.mpg)

Call:
lm(formula = price ~ city.mpg)

Coefficients:
(Intercept)     city.mpg
    35947.4       -894.8
```

This implies that the price of the car *decreases* by a factor of \$ -894.80 per MPG. This interpretation is misleading; in reality, more powerful cars typically have much lower mileage than less powerful cars (Fig. 6.4). As the predicted prices for cars increase with more horsepowers, we have the effect of prices decreasing for cars with high mileage. The multivariate regression analysis of Eq. 6.7 shows the relationship between car price, horsepower, and mileage considered together. From the coefficient of the `city.mpg` variable, we observe that the car price increases by \$ 61.51 per MPG while controlling for horsepower, engine size, and length variables. Among cars of the same power and size, the cars with higher mileage are likely to be more expensive.

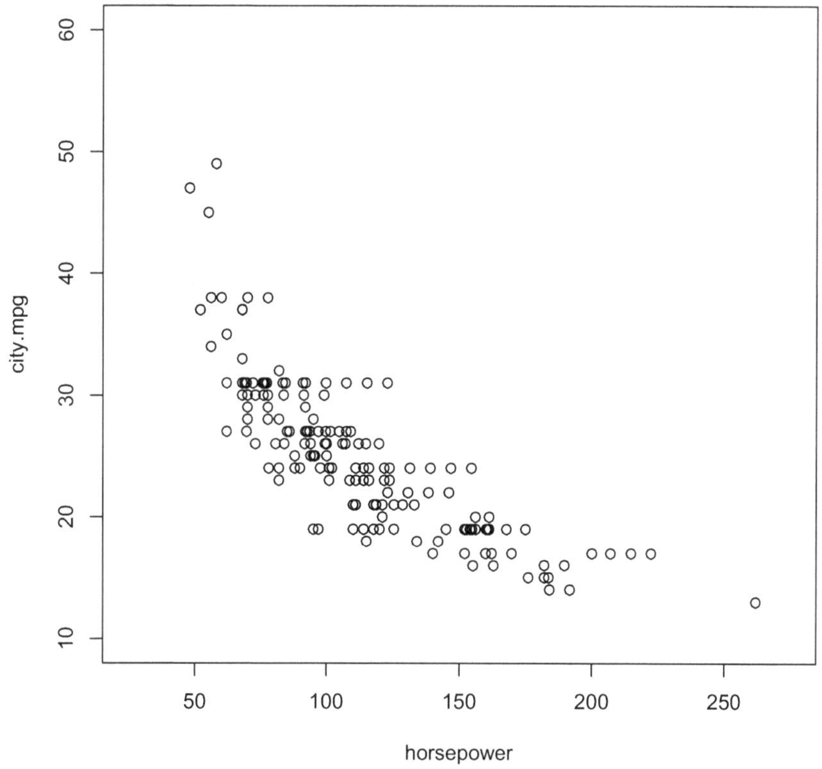

Fig. 6.4 City MPG vs. horsepower

6.2.3 *Log-Linear Regression Models*

A linear regression model makes a modeling assumption about the linear relationship between variables. This assumption is not a problem when the variables have a linear or close to linear relationship. If the data has a nonlinear trend, a linear regression model cannot fit the data accurately. One approach for modeling nonlinear data is to fit a model for a transformed variable. A commonly used transformation is to use the log() function on the variables.

The log-linear regression models are not fundamentally different from their linear counterparts. We use the logarithm of our variable as the predictor variable and then perform linear regression. We reuse the `lm()` function, as the model coefficient computation is also not different for linear and log-linear models. The transformation can be applied on the target variable or one or more of the predictor variables.

A good example of variables with nonlinear relationship is `city.mpg` vs. `horsepower` as shown in Fig. 6.4. In general, we see that the car mileage decreases for increasing values of horsepower. The decrease, however, is much steeper for smaller values of horsepowers and is much more gradual for larger values of horsepowers. In a linear relationship, the rate of decrease would be similar across all values of horsepower. Instead of performing linear regression between `city.mpg` and `horsepower`, we use a log transformation on the horsepower variable. We could have applied a log transformation on `city.mpg` as well; in this case we choose to apply it on `horsepower` because, `city.mpg` has a decreasing relationship with `horsepower`. In the following code, we call `lm()` with the formula `city.mpg ~ log(horsepower)`.

```
> lm(city.mpg ~ log(horsepower))

Call:
lm(formula = city.mpg ~ log(horsepower))

Coefficients:
    (Intercept)  log(horsepower)
         101.44           -16.62
```

Here the relationship between the variables is:

$$\text{city.mpg} = -16.62 \times \log(\text{horsepower}) + 101.44.$$

The negative coefficient indicates that `city.mpg` decreases for increasing values of `horsepower`.

We compare the model fit of the log-linear regression model with the linear regression model. Using the summary() function, we observe that the linear regression model has a residual standard error of 3.538.

```
> summary(lm(city.mpg ~ horsepower))

Call:
lm(formula = city.mpg ~ horsepower)

Residuals:
    Min      1Q  Median      3Q     Max
-7.5162 -1.9232 -0.1454  0.8365 17.2934

Coefficients:
             Estimate Std. Error t value Pr(>|t|)
(Intercept) 39.842721   0.741080   53.76   <2e-16 ***
horsepower  -0.140279   0.006725  -20.86   <2e-16 ***
---
Signif. codes:  0 *** 0.001 ** 0.01 * 0.05 . 0.1   1

Residual standard error: 3.538 on 191 degrees of freedom
Multiple R-squared: 0.6949, Adjusted R-squared: 0.6933
F-statistic: 435.1 on 1 and 191 DF,  p-value: < 2.2e-16
```

The log-linear model has a residual standard error of 2.954 indicating that it fits the data better than the linear model.

```
> summary(lm(city.mpg ~ log(horsepower)))

Call:
lm(formula = city.mpg ~ log(horsepower))

Residuals:
    Min      1Q  Median      3Q     Max
-6.7491 -1.7312 -0.1621  1.2798 15.0499

Coefficients:
                Estimate Std. Error t value Pr(>|t|)
(Intercept)     101.4362     2.8703   35.34   <2e-16 ***
log(horsepower) -16.6204     0.6251  -26.59   <2e-16 ***
---
Signif. codes:  0 *** 0.001 ** 0.01 * 0.05 . 0.1   1

Residual standard error: 2.954 on 191 degrees of freedom
Multiple R-squared: 0.7873, Adjusted R-squared: 0.7862
F-statistic:   707 on 1 and 191 DF,  p-value: < 2.2e-16
```

We can also observe how well the log-linear model fits the data visually. In Fig. 6.5, we show the log-linear model by a solid line and the linear model by a dashed line. As the linear model assumes a linear relationship between `city.mpg` and `horsepower`, its regression function is a straight line. Due to the log transformation, the log-linear model has a curved regression function. We can see that this curve better represents the data, especially for the small and large values of the `horsepower` variable.

6.3 Nonparametric Regression Models

A log-linear regression model provides a way to represent more complex relationships between variables than a linear regression model. This model still makes an assumption though: target variable has a log-linear relationship to the predictor variables. For many datasets, such a relationship might not hold true. In fact, there might not be a well-defined relationship between variables which we can represent using a function transformation. In this section we look at regression models that do not make any assumptions about the underlying data.

The class of techniques that make explicit assumptions about the model structure are called parametric models or in general parametric statistics. An alternative method of creating models that does not involve making any implicit or explicit assumptions about the data is called nonparametric modeling or in general nonparametric statistics. In nonparametric regression, we consider the data itself to be the model; the regression function we learn will be specific to the shape of the given dataset.

The nonparametric approach has a few pros and cons. On one hand we can fit a complex regression function that is more aligned with the the data: without making any assumptions on the shape of the function. On the other hand, using nonparametric

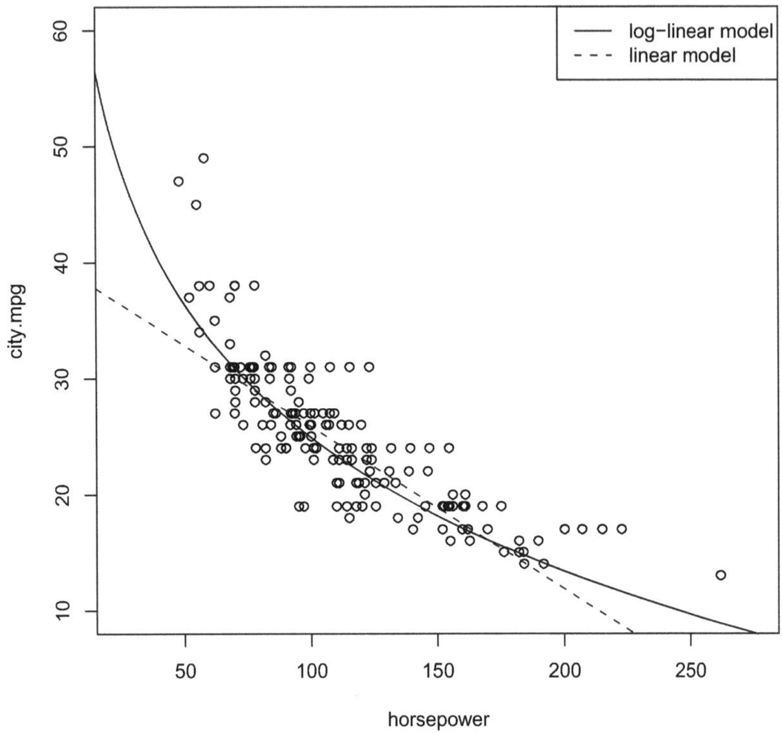

Fig. 6.5 Log-linear vs. linear regression model

techniques, we might fit the data *too well*. We can view a given dataset as a sample drawn from the data distribution: e.g., the daily temperature of a city in last year is a sample of the general weather or the climate of that city. What we ultimately want to learn is a representation for the data distribution, not the given dataset. A dataset may have biases or noise, e.g., if the the city had unusually cold weather for a few days that year, we might always make consistently lower temperature predictions than it is usual for those days. When a model fits to the noise in the dataset as opposed to the relationship between variables, it is said to *overfit* the data. As a consequence, we usually need a comparatively larger dataset to fit an accurate nonparametric model.

There are many approaches for nonparametric regression; we look at a few below.

6.3.1 Locally Weighted Regression

In locally weighted regression (LWR), we combine the outputs of multiple regression models to obtain the regression function. The local aspect of this model is from how we fit the individual regression models. For each data point x, we fit a weighted regression model. We assign higher weights to the data points near to the given data point x. The value of the regression function at x is the predicted value of the weighted regression model at x. The regression function for the LWR model is the set of predicted values that we obtain for every data point in the dataset. As the individual regression models are dependent on the data points, an LWR model is aligned with the shape of the data. For this reason, LWR is also called *smoothing*, as it replaces data points with their predicted values that are computed using their nearby data points. We also use LWR in scatter plots to highlight the relationship between variables.

LWR is a nonparametric approach as it does not impose any restriction on the shape of the global regression function: linear or otherwise, even though we may be using parametric approaches for the local regression models.

LWR as we described above is fairly general; we can use any type of regression model for the local fitting. Also, there are different schemes for assigning weights to the data points depending on their distance from the candidate data point. A widely used LWR scheme is LOESS [2], where we use linear or quadratic regression models for local fitting and a tri-cube weight function. LOESS is the generalization of the older LOWESS scheme that only used linear regression for local fitting.

Another parameter that we use in LOESS is the span α, which is the proportion of the data points that we use for local fitting, i.e., for a dataset with n data points, we only use a subset of αn data points that are nearest to the candidate data point x. This makes LOESS a nearest-neighbor approach: it is computationally much more expensive than linear regression, as we need to evaluate multiple local regression models.

If $d(x)$ is the distance between the candidate data point and the furthest data point in the subset, the weight for a data point x_i is given by

$$w(x, x_i) = \left(1 - \left|\frac{x - x_i}{d(x)}\right|^3\right)^3. \tag{6.8}$$

From this formula, we see that the candidate data point x has the highest weight of one, and the furthest data point has zero weight. Due to the cube functions, the weights for the data points decrease rapidly with their distance from x.

In R, we use the `loess()` function for computing a LOESS model. This function also takes a formula as input similar to the `lm()` function. We fit a LOESS model for car mileage and horsepower below.

```
> loess(city.mpg ~ horsepower)
Call:
loess(formula = city.mpg ~ horsepower)

Number of Observations: 193
Equivalent Number of Parameters: 5.37
Residual Standard Error: 2.673
```

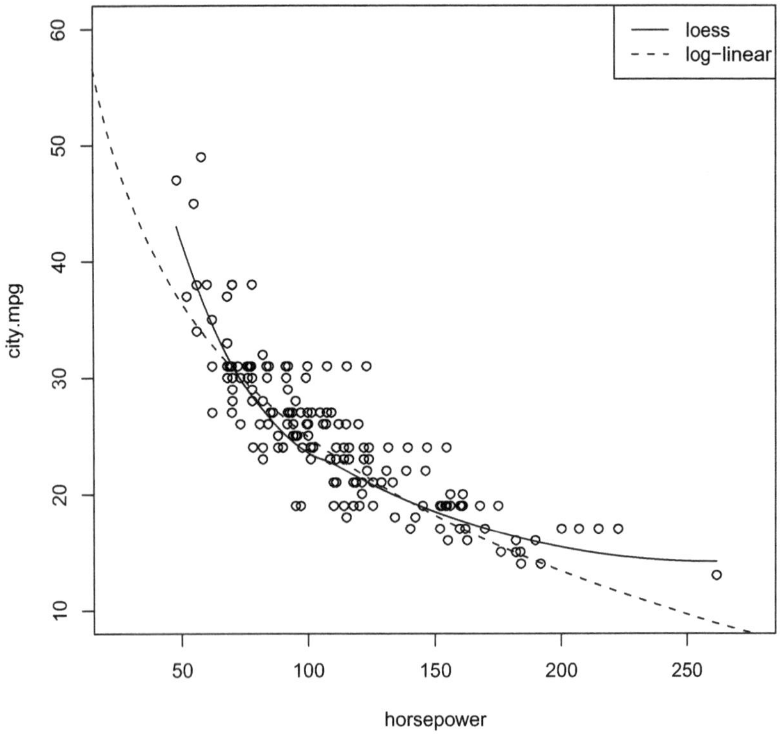

Fig. 6.6 LOESS vs. log-linear regression model

The residual standard error of 2.673 indicates that LOESS fits the data better for `city.mpg` and `horsepower` than both linear and log-linear regression models. This becomes more apparent when we visualize the model fits in Fig. 6.6. The solid line indicates the regression function curve for LOESS and the dashed line indicates that for the log-linear model. We see that the shape of the regression curve for LOESS changes with the data: much sharper for smaller values of `horsepower` and flatter for larger values. As the LOESS model is a combination of multiple locally fitted regression functions, we see kinks in the function such as the one near `horsepower` = 100.

By default, `loess()` uses a span of 0.75 and second-degree (quadratic) polynomials for the local regression models. We can configure these using the `span` and `degree` parameters. We show the regression function curves for different span values in Fig. 6.7. The curve for `span = 0.21` has much sharper kinks as compared to the curve for `span = 0.75`. This is because smaller spans imply smaller neighborhoods for fitting the local regression models, which means we end up with more disparate local regression models for nearby data points. Due to this we have variance in the predicted output for the consecutive local regression models, which in turn causes kinks in the regression curve.

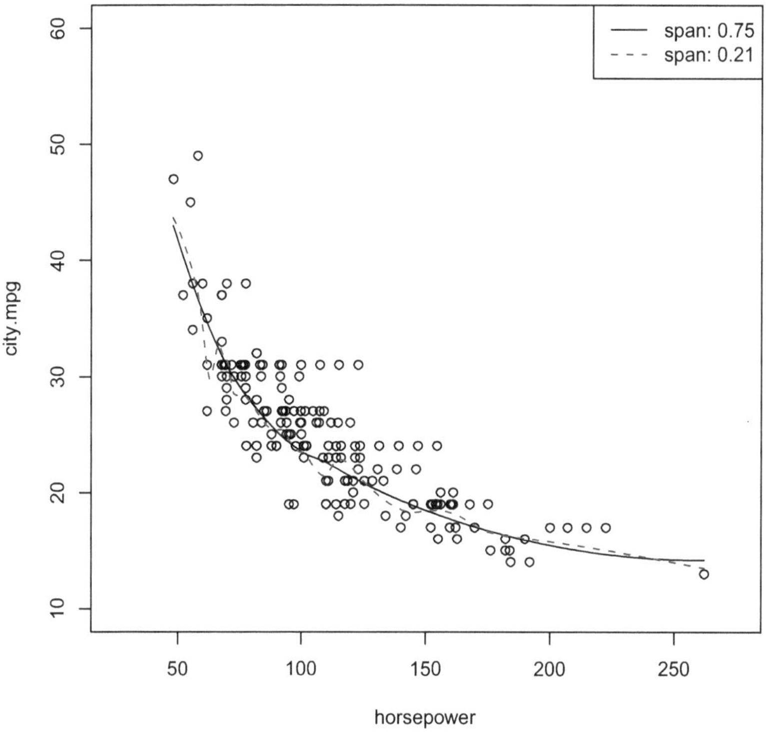

Fig. 6.7 LOESS vs. log-linear regression model

Using an appropriate span is important: having a very small span causes the regression function to be simply predict the target variable at that data point. Using a large span causes oversmoothing, where the neighborhood for local fitting is so large that we end up averaging out the variances in the data points.

We can also use `loess()` for multivariate regression. Similar to `lm()`, we only need to specify the predictor and target variables in the formula.

6.3.2 *Kernel Regression*

Kernel regression[6] is another nonparametric approach where we compute the value of the predictor variable at each data point by taking a weighted average of the target variable at all data points. The weights are given by the kernel function, which we can think of as a measure of distance between two data points. The data points nearer to the candidate data point have high weight and those further away from the data

[6] Kernel regression is also called Nadaraya–Watson kernel regression, due to its coinventors.

point have low weight. The rate of decrease of weight is controlled by the bandwidth parameter. For the data points $x_1, \ldots, x_n$ with target values $y_1, \ldots, y_n$, the weighted average value at the data point x is given by

$$f(x) = \frac{\sum_{i=1}^{n} K(x, x_i) y_i}{\sum_{i=1}^{n} K(x, x_i)}.$$

In this model, the regression function is the set of values of the predictor variable at each data point. Kernel regression is similar to LOESS as both methods use weights to compute local predictions. There are important differences though: in kernel regression, we use the weighted average to obtain the predicted value, while in LOESS we fit a local model at each candidate data point and obtain the predicted value from this model.

There are many kernel functions used in practice. We list a few popular functions below with h as the bandwidth parameter.

1. Gaussian kernel:

$$K(x, x_i) = \frac{1}{\sqrt{2\pi}} e^{-\frac{(x-x_i)^2}{2h^2}}.$$

2. Uniform kernel:

$$K(x, x_i) = \frac{1}{2} I\left(\left|\frac{x - x_i}{h}\right| \leq 1\right).$$

3. Triangular kernel:

$$K(x, x_i) = \left(1 - \left|\frac{x - x_i}{h}\right|\right) I\left(\left|\frac{x - x_i}{h}\right| \leq 1\right).$$

4. Epanechnikov kernel:

$$K(x, x_i) = \frac{3}{4}\left(1 - \left(\frac{x - x_i}{h}\right)^2\right) I\left(\left|\frac{x - x_i}{h}\right| \leq 1\right).$$

The `np` package in R provides an implementation of kernel regression. We use the `npreg()` function to fit a kernel regression model. Similar to `lm()`, this function also takes a formula denoting the predictor and target variables as input. We fit a kernel regression model for car mileage and horsepower below.

```
> npreg(city.mpg ~ horsepower)

Regression Data: 193 training points, in 1 variable(s)
                 horsepower
Bandwidth(s):    1.462226

Kernel Regression Estimator: Local-Constant
```

```
Bandwidth Type: Fixed

Continuous Kernel Type: Second-Order Gaussian
No. Continuous Explanatory Vars.: 1
```

By default, `npreg()` uses the Gaussian kernel function. We can specify a different kernel function using the `kernel` parameter.

Also, `npreg()` automatically selects the bandwidth using crossvalidation as well. In the above example the model used a bandwidth of 1.4622. We obtain the other model statistics using the `summary()` function.

```
> summary(model.np)

Regression Data: 193 training points, in 1 variable(s)
                horsepower
Bandwidth(s):    1.462226

Kernel Regression Estimator: Local-Constant
Bandwidth Type: Fixed
Residual standard error: 2.165175
R-squared: 0.8848329

Continuous Kernel Type: Second-Order Gaussian
No. Continuous Explanatory Vars.: 1
```

We observe that the kernel regression model has a residual standard error of 2.1652 which is better than linear, log-linear, and LOESS regression models for the same data. We visualize the regression function in Fig. 6.8.

We can also use `npreg()` to perform multivariate regression as well. Similar to `lm()`, we only need to specify the predictor and target variables in the formula.

6.3.3 Regression Trees

Decision trees are one of the most widely used models in all of machine learning and data mining. A tree is a data structure where we have a root node at the top and a set of nodes as its children. The child nodes can also have their own children, or be terminal nodes in which case they are called leaf nodes. A tree has a recursive structure as any of its node is the root of a subtree comprising the node's children. Binary tree is a special type of tree where every node has at most two children.

We construct decision trees as a model for our data. In a decision tree, we start with the complete dataset at the root. At each node, we divide the dataset by a variable; the children of a node contain the different subsets of the data. We assign the mean value of the target variable for the data subset at a leaf node, which is our prediction.

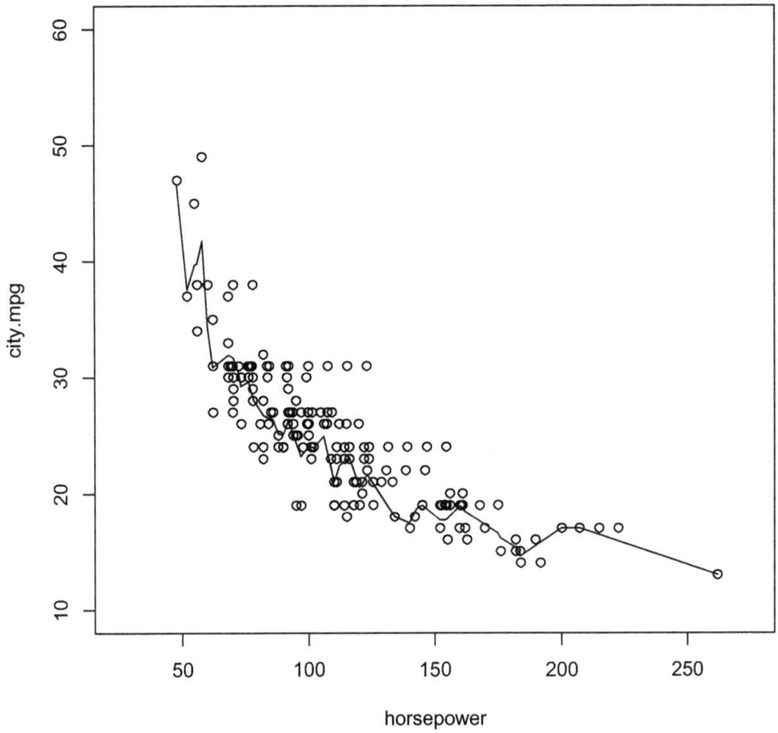

Fig. 6.8 Kernel regression

We can use a decision tree for either regression and classification, in which case we refer to it as a regression tree or a classification tree.

As an example, let us look at a decision tree modeling the survival rates of the passengers aboard the Titanic shown in Fig. 6.9.[7] Initially, we start with the entire list of passengers. At the root node, we first divide the data by splitting on the `sex` variable: the male passengers go to the left subtree and the female passengers go to the right subtree. In the right subtree, we further divide the female passengers by class: the passengers in the third class go to the left subtree and passengers in the first and second classes go to the right subtree. The numbers in the leaf nodes denote the survival rates for that group of passengers; we see that only 19.1 % of male passengers survived, while 49.07 % of female passengers in the third class, and 93.2 % of female passengers in the first and second classes survived.

There are many advantages of using decision trees. Firstly, they provide a visual representation of the data. This is especially useful for a nontechnical audience.

[7] This decision tree is created using the passenger data obtained from the Encyclopedia Titanica [3].

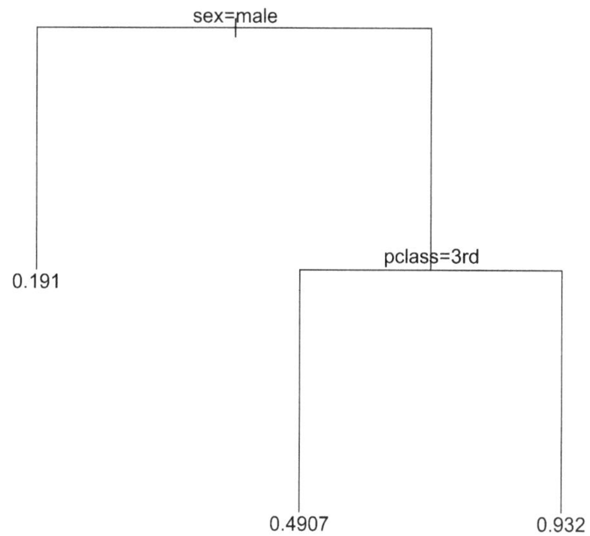

Fig. 6.9 Decision tree for survival rates of passengers aboard the Titanic

Secondly, decision trees do not require as much data preprocessing as compared to other methods.

When we are using decision trees as our regression model, the problem comes down to finding the best tree that represents our data. Unfortunately, finding the optimal decision tree, i.e., the tree that minimizes the residual standard error or any other model fit metric, is computationally infeasible.[8] To find an optimal decision tree, we would need to evaluate all possible tree configurations consisting of all combinations of variable splits at every node. This combinatorial expansion is not practical for a dataset with a nontrivial number of variables. In practice, we use a greedy strategy for efficiently constructing a decision tree. In this strategy, at every level of the tree, we choose a variable that splits the data according to a heuristic. Intuitively, we require that the two data splits are homogeneous with respect to the target variable. For instance, in the Titanic dataset, we chose to split on the class variable for female passengers as the survival rates of first and second class passengers are similar to each other compared to the survival rate for third class passengers.

Some of the commonly used heuristics for creating decision trees include:

1. Information gain: used in the C4.5 algorithm
2. Gini coefficient: used in classification and regression tree (CART) algorithm

We use the CART algorithm to create decision trees. In R, this algorithm is implemented in the `rpart` package. The `rpart()` function takes a formula denoting the predictor and target variable as input and generates the corresponding decision tree.

[8] The computational complexity of finding the optimal decision tree is NP-hard.

We fit a decision tree for price of the car given its length, engine size, horsepower, and mileage below.

```
> fit = rpart(price ~ length + engine.size
              + horsepower + city.mpg)
> fit
n= 193

node), split, n, deviance, yval
      * denotes terminal node

 1) root 193 12563190000 13285.030
   2) engine.size< 182 176  3805169000 11241.450
     4) horsepower< 94.5 94   382629400  7997.319
       8) length< 172.5 72   108629400  7275.847 *
       9) length>=172.5 22   113868600 10358.500 *
     5) horsepower>=94.5 82  1299182000 14960.330
      10) length< 176.4 33   444818200 12290.670
        20) city.mpg>=22 21    94343020 10199.330 *
        21) city.mpg< 22 12    97895460 15950.500 *
      11) length>=176.4 49   460773500 16758.270 *
   3) engine.size>=182 17   413464300 34442.060 *
```

The rpart() function prints the model in a textual form. We display the decision tree as a figure using the plot() function. We display the node labels using the text() function. Figure 6.10 shows the output.

```
> plot(fit,uniform=T)
> text(fit,digits=6)
```

We set the `uniform` and `digits` parameters for pretty printing.

One problem with a greedy strategy is that sometimes it creates a complex tree with a large number of data splits. Due to this, the subset of data corresponding to a leaf node is very small, which might lead to inaccurate predictions. This is equivalent to model overfitting; to avoid this, we use a pruning strategy to remove unnecessary branches of the tree.

We prune the decision tree using the `prune()` function of the `rpart` package and specify the level of pruning using the complexity parameter `cp`. Setting `cp` = `1` prunes all nodes of the tree up to the root, while setting `cp` = `0` does not prune the tree at all. We prune the decision tree we obtained above using the parameter `cp` = `0.1` and display it in Fig. 6.11. We observe that the decision tree has relatively fewer branches.

```
> fit.prune = prune(fit,cp=0.1)
```

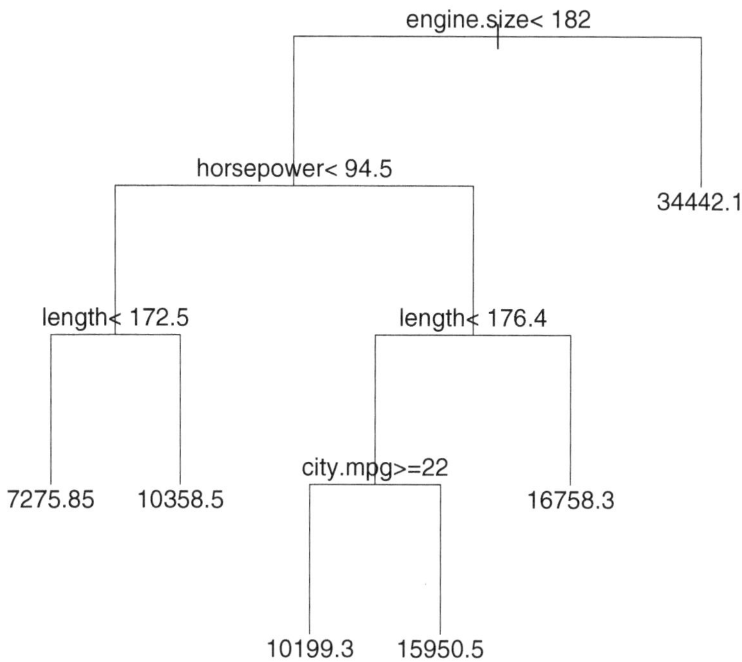

Fig. 6.10 Decision tree for predicting the price of a car

Fig. 6.11 Pruned decision tree for predicting the price of a car

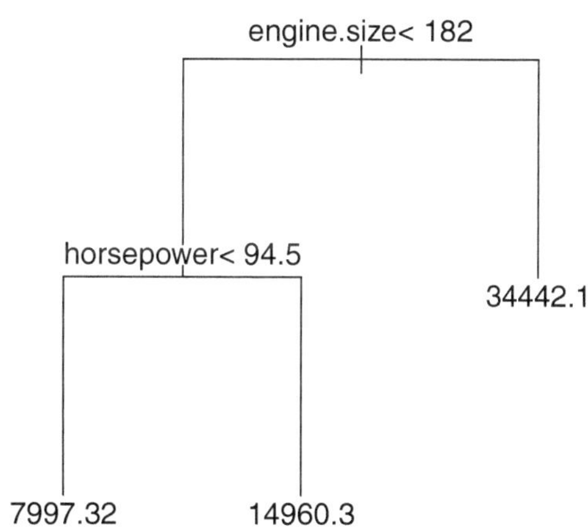

Apart from CART, there are many other algorithms for creating regression trees. There also exist algorithms that combine the output of multiple regression trees. One powerful example of this approach is the random forest algorithm available in R through the `randomForest` package.

6.4 Chapter Summary

In this chapter, we looked at regression analysis which is one of the most important and commonly used data science tasks. A regression model is used to predict a numeric target or dependent variable using one or multiple predictor predictor or independent variables. Regression models are classified into two types: parametric, where we assume a particular type of relationship between the variables and non-parametric, where we do not explicitly assume such a relationship. In parametric models, we looked at simple or single-variable and multivariate linear regression where we assume a linear relationship between variables. Similarly, a log-linear model assumes a log-linear relationship. In R, we use the `lm()` function in the base package to fit these models and `predict()` function to obtain model predictions.

In nonparametric models, we looked at LWR models which combine the output of multiple linear regression models which could be parametric or nonparametric. A popular LWR model is LOESS that uses local linear or quadratic regression models. Kernel regression is another nonparametric model which uses the weighted average of neighboring data points as predictor. We use the `npreg()` function to obtain kernel regression models. Finally, we looked at decision tree models for regression which learn a tree structure from the data. Thus, at every node of the tree, we split the dataset by a variable and make predictions by averaging the target variable values at the leaf node. We use the rpart package to fit the decision trees using the CART algorithm.

References

1. Top 10 least powerful cars on sale in the U.S. http://wot.motortrend.com/top-10-least-powerful-cars-on-sale-in-the-united-states-242699.html#axzz2T8brud72. Accessed 1 Aug 2014.
2. Cleveland, W. S. (1979). Robust locally weighted regression and smoothing scatterplots. *Journal of the American Statistical Association, 74,* 829–836.
3. http://www.encyclopedia-titanica.org/. Accessed 1 Aug 2014.

Chapter 7
Classification

7.1 Introduction

Along with regression, classification is an important modeling technique in the data science toolset. Instead of predicting a numeric value as we did in regression, we fit a classification model to *classify* data points into multiple categories or *classes*. In this case, we also refer to the target variable as the class label. Binary classification is a special type of classification with the class label taking two values. Classification models with target variables taking more than two values is called multiclass classification.

Classification is also one of the central problems in machine learning. It is widely used in practice; classification finds application in problems such as spam filtering, computational advertising, speech and handwriting recognition, and biometric authentication, to name a few.

There are a large number of approaches for classification; in this chapter, we will look at some of the commonly used ones. Each classification model has a different intuition behind it. It is important to know this intuition when choosing the right classifier for our data. In this chapter we will look at how each classification model internally works to gain this understanding.

Similar to Chap. 6, we will study both parametric and nonparametric approaches. In parametric models, we first look at naive Bayes (NB), which is perhaps the simplest of all classification models. In this model, we compute probability of a data point belonging to a class or the class probability using the *Bayes theorem*. Next we will look at logistic regression, which is another classification model, where we model the class probability using a *sigmoid function*. Finally, we look at the more sophisticated support vector machine (SVM) classifiers. In SVMs, we use the concept of *margin*, which is the minimum distance between the data points of the two classes as given by the classifier. The classifier obtained by SVM is the one that maximizes the margin. We will also look at using SVMs with kernels for classifying datasets with classes that are not linearly separable.

In nonparametric models, we look at nearest neighbors and decision tree-based models for classification. These models are similar to the nonparametric models for regression that we saw in Section 6.3.

M. A. Pathak, *Beginning Data Science with R*,
DOI 10.1007/978-3-319-12066-9_7

In R, we rarely need to implement the classification models from scratch. In this chapter, we will look at R packages that provide functions to fit and predict the classification models mentioned above.

Case Study: Credit Approval
In this chapter, we look at a dataset related to the credit approvals for the customers of a German bank.[1] The dataset contains different attributes of an individual such as the purpose for the loan, prior credit history, employment, along with the information about whether the loan was approved for that individual. Table 7.1 contains the list of variables used in this dataset. The credit request approval variable is the class label; we train a classifier that determines whether to approve the credit for a new individual. This is similar to what a financial institution such as a bank or credit card company does when evaluating the risks of each loan application. In terms of descriptive analysis, the parameters of the classification model also provide insight into which attribute of the individual is important for credit approval.

We load and attach the dataset in the code below.

```
> data = read.csv('credit.csv')
> attach(data)
```

7.1.1 Training and Test Datasets

When developing a classifier, we usually split the dataset into training and test datasets. We use the training data to fit the model and test data to evaluate it. This is because it is generally not a good idea to evaluate a model on the same data it is trained on. Having a model that fits the training data with very high accuracy indicates that it is also fitted to some of the biases in that dataset. Out goal is to obtain a model that is able to predict new data points. We compare the performance of two models by their accuracy on the test data.[2]

Our dataset has 1000 data points; we assign the first 750 data points to the training data and remaining 250 data points to the test data.[3]

```
> train = data[1:750,]
> test = data[751:1000,]
```

[1] Data obtained from the UCI Machine Learning repository. http://archive.ics.uci.edu/ml/datasets/Statlog+(German+Credit+Data)

[2] Having a test dataset is better than evaluating on training data alone, but is not ideal. In the process of optimising our model to minimize the classification error on test data, we might fit to the biases in this dataset. Techiniques such as crossvalidation are designed to solve this problem [3].

[3] We chose a deterministic train/test split so that the results in this chapter are reproducible. In practice, the train/test split should be uniformly random.

Table 7.1 List of variables in automobiles dataset

Variable name	Description
over.draft	Over-draft status of the existing checking account
credit.usage	Credit usage duration in months
credit.history	Credit history
purpose	Purpose of the loan
current.balance	Current balance
average.credit.balance	Credit amount
employment	Present employment period
Location	Location code
personal.status	Marital status and sex
other.parties	Other debtors/guarantors
residence.since	Present residence since
property.magnitude	Property held
cc.age	Age in years
other.payment.plans	Other payment plans
housing	Type of housing
existing.credits	Existing loans
job	Type of job
num.dependents	Number of dependents
own.telephone	Does the individual own a telephone
foreign.worker	Is the individual worker a foreign national
class	Whether the credit request was approved

7.2 Parametric Classification Models

7.2.1 Naive Bayes

Naive Bayes (NB) is a probabilistic model for classification. The probablistic aspect of this model is that it computes the probability of a data point being in either of the classes. This method is based on Bayes theorem, which is also the foundation of *Bayesian statistics*.[4] Bayesian statistics is based on the notion of a prior probability, which is our prior belief about the world before observing the data. The posterior

[4] Bayesian statistics and its analogue frequentist statistics are the two main families of statistical analysis techniques that are based on different principles. The superiority of each family over the other is often debated in academic circles, but statistical techniques from both families find abundant practical applications.

probability on the other hand, is the our updated information about the world taking into account what we saw in the data. The Bayes theorem states that posterior probability is proportional to the data likelihood times the prior probability. If $P(\theta)$ is our prior probability, $P(x|\theta)$ is the likelihood of the data x, the then the posterior probability $P(\theta|x)$ is given as:

$$P(\theta|x) = \frac{P(x|\theta)P(\theta)}{P(x)}.$$

In a naive Bayes classifier, we compute the conditional probability of a data point having a class label. Let us consider a binary classification setting, with the class label $y = \{1, -1\}$ and the data point x. Using the Bayes theorem,

$$P(y = 1|x) = \frac{P(x|y = 1)P(y = 1)}{P(x)},$$

and

$$P(y = -1|x) = \frac{P(x|y = -1)P(y = -1)}{P(x)}.$$

We can assign the label $y = 1$ or $y = -1$ to the data point x depending on which posterior probability is greater. The denominator $P(x)$ does not play a role as it is the same for the both posterior probabilities.

All we need to do is use the training data to compute the probabilities $P(y = 1)$, $P(y = -1)$, $P(x|y = 1)$, and $P(x|y = -1)$ and we will have our classifier. The former two are straightforward: $P(y = 1)$ is the fraction of training data points with class label 1, and analogously, $P(y = -1)$ is the fraction of training data points with class label -1. These are called marginal probabilities or simply *marginals*. Computing $P(x|y = 1)$ is tricky; we need to count how many times the same data point x appeared in the training data with labels 1 and -1. We should be able to do so for a single input data point x. This might even be possible if we represent a data point using only a few features taking a small set of values. In most nontrivial settings, however, we would not be able to compute these probabilities unless we have enormous amounts of data for all feature combinations.

To solve this problem of computing $P(x|y = 1)$, the NB model makes an assumption that the features of a data point are independent given its label. For the data point x with features $x_1, x_2, \ldots, x_d$, we use conditional independence to compute the joint probability $P(x|y = 1)$ as the product of individual conditional probabilities.

$$P(x|y = 1) = P(x_1, x_2, \ldots, x_d|y = 1) = P(x_1|y = 1) \cdot P(x_2|y = 1) \cdots P(x_d|y = 1).$$

Computing the feature given class probability $P(x_j|y)$ is more reasonable; it is the fraction of data points with that feature value x_j and class label y. In many practical settings, there are often at least a few data points that share a common feature.

Although an NB model does provide estimates for class probabilities, it is infrequently used as a method for density estimation. Due to the naive assumption, the probability estimates are not accurate enough by themselves, but are sufficiently accurate to determine which class has a greater probability.

One of the first major applications of NB was spam filtering [4]. In this setting, text documents such as emails are treated as a data point x with a bag of words forming the features $x_1, \ldots, x_d$. The class labels are given by whether a document is marked as spam or not. Identifying the class conditional probabilities is simply identifying the distribution of spam vs nonspam documents for each word.

7.2.1.1 Training an NB Classifier Using the e1071 Package

In the example below, we train a classifier to predict the `class` variable in the credit dataset.

To start off, let us consider only the `housing` variable. `housing` is a categorical variable that takes three values: either the individual owns a house, lives in rental accomodation, or lives for free. Our intuition is that individuals living in their own house are likely to have good credit as they also have sufficient collateral to cover their debts.

We use the `table()` function to compute the distribution of `housing`.

```
> table(train$housing)

for free      own     rent
      83      537      130
```

There are 537 individuals living in their own houses, as opposed to 130 individuals living in rented houses, and 83 individuals living for free.

Let us look at the fraction of these individuals having good or bad credit.

```
> table(train$class,train$housing)

       for free own rent
  bad        31 141   51
  good       52 396   79
```

We see that only 141 out of 537 or 26.26 % individuals living in their own houses have bad credit, as opposed to 37.35 % and 39.23 % for individuals living for free and for rent, respectively. Similarly, we obtain the distribution of the class variable below.

```
> table(train$class)

 bad good
 223  527
```

We see that 223 out of 750 or 29.73 % individuals have bad credit and 527 out of 750 or 70.27% individuals have good credit. Similarly, we can compute the class-conditional probabilities, e.g., the probability of an individual living for free with bad credit is $31/223 = 13.90\,\%$.

An NB model consists of such marginal and class-conditional probabilities computed for all feature value/class pairs. As with most statistical algorithms, we do not need to write code for this in R; we use the `naiveBayes()` function provided by the `e1071` package. This function follows a similar convention as the `lm()` function: it also takes a formula denoting the predictor and target variables as input along with the data frame containing the training data.

We train a model using all features below.

```
> library(e1071)
> model.nb = naiveBayes(class ~ ., train)
```

We observe the model probabilities by printing the model object.

```
> model.nb

Naive Bayes Classifier for Discrete Predictors

Call:
naiveBayes.default(x = X, y = Y, laplace = laplace)

A-priori probabilities:
Y
      bad      good
0.2973333 0.7026667

Conditional probabilities:
      over.draft
Y              <0      >=200   0<=X<200
  bad  0.40358744 0.04932735 0.39910314
  good 0.20113852 0.07590133 0.23719165
      over.draft
Y       no checking
  bad    0.14798206
  good   0.48576850

... (output truncated for brevity)

      housing
Y         for free        own       rent
  bad  0.13901345 0.63228700 0.22869955
  good 0.09867173 0.75142315 0.14990512

      existing.credits
Y          [,1]      [,2]
  bad  1.372197 0.5699203
  good 1.400380 0.5660213
```

```
        job
Y          high qualif/self emp/mgmt    skilled
  bad                     0.18834081 0.59641256
  good                    0.13472486 0.64326376
        job
Y          unemp/unskilled non res unskilled resident
  bad                   0.02242152          0.19282511
  good                  0.01707780          0.20493359

        num.dependents
Y              [,1]      [,2]
  bad  1.152466 0.3602811
  good 1.148008 0.3554449

        own.telephone
Y              none       yes
  bad  0.6278027 0.3721973
  good 0.5958254 0.4041746

        foreign.worker
Y                no         yes
  bad  0.00896861 0.99103139
  good 0.04554080 0.95445920
```

The class probabilities for the `housing` variable given by the model are the same as the ones we began computing ourselves above.

Similar to `lm()`, we use the `predict()` function to predict the class labels of the test dataset using this model. This function takes the model object and an optional data frame as input. If the input data frame is not provided, the function uses the data frame used to train the model. The output of this function is a factor containing the class values for the respective data point.

```
> predict(model.nb,test)
   [1] good bad  good good good bad  good good good
  [10] good good good good bad  good good bad  good
...
[244] bad  good good good good bad  good
Levels: bad good
```

We compute the accuracy of the classifier by comparing the predicted labels with the actual labels of the test dataset.

```
> table(predict(model.nb,test) == test$class)/length
                                  (test$class)

FALSE  TRUE
0.208 0.792
```

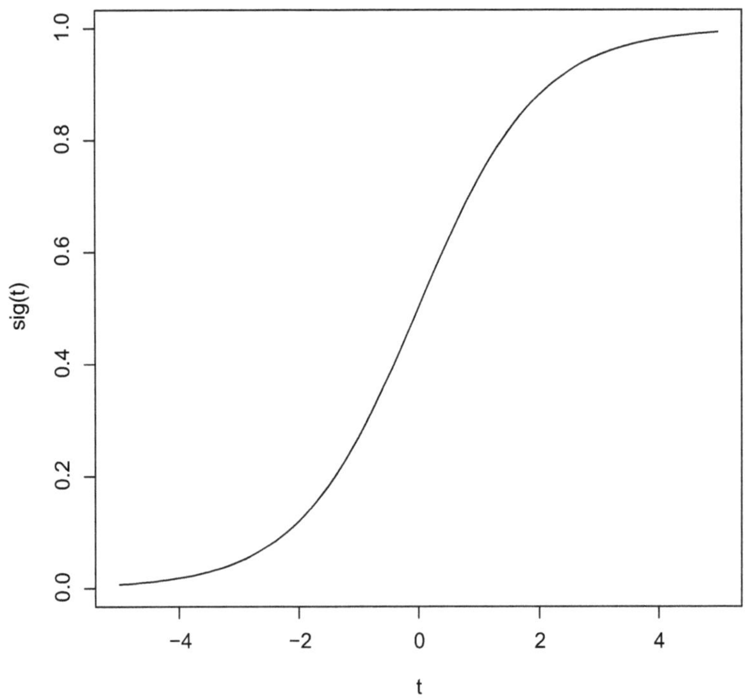

Fig. 7.1 Sigmoid function

The accuracy of our naive Bayes classifier `model.nb` is 79.2 %.

7.2.2 *Logistic Regression*

Logistic regression is the workhorse of classification algorithms. It is widely used in almost all classification applications involving both categorical and numeric data. Despite its name, logistic regression is a method for binary classification. The building block of a logistic regression model is the logistic or *sigmoid* function defined as:

$$sig(t) = \frac{1}{1 + e^{-t}}. \tag{7.1}$$

It is called the sigmoid function because of it is shaped like the letter "S." Figure 7.1 shows the plot of this function. We observe two things about this function: firstly, the range of the function is 0–1. Secondly, it is a monotonically increasing function, which means $sig(t_1) > sig(t_2)$ for all $t_1 > t_2$. These two properties make the sigmoid function convenient to model probabilities.

Logistic regression is also a probabilistic model of classification. In a logistic regression model, we fit the conditional probability of a data point x belonging to class y: $P(y|x)$ using the sigmoid function. If the class label y takes two values $+1$ and -1, and the data point has only one predictor variable, we have

$$P(y = 1|x) = sig(w_0 + w_1 x) = \frac{1}{1 + e^{-(w_0 + w_1 x)}}. \tag{7.2}$$

The probability for the class label being -1 is simply the complement of the above probability. Using algebraic manipulations, we can see that it is equal to $sig(-(w_0 + w_1 x))$. This allows us to represent the probabilities of the class label y being either $+1$ or -1 compactly as:

$$P(y|x) = sig(y(w_0 + w_1 x)). \tag{7.3}$$

Given a dataset, there are many possible criteria for selecting the best coefficient vector. In a logistic regression model, we select the coefficient vector w that maximizes the log-likelihood, which is the joint probability of the data points in the training data having their corresponding label. Given this log-likelihood function, we obtain w using optimization techniques such as gradient descent.

7.2.2.1 Using the glm() Function

We fit a logistic regression classification model in R using the `glm()` function, which is used to fit generalized linear model. As the name suggests, glm is a generalization of the linear regression model (Sect. 6.2.1). Apart from the formula denoting the predictor and target variables, `glm()` also takes a link function as its argument. The link function determines the probabilistic relationship between the target and the predictor variables. We obtain regression or classification models by choosing different link functions; notable examples include the Gaussian distribution which gives us a linear regression model, and binomial distribution which gives us a logistic regression model.

In the `glm()` function, we specify the distribution using the `family` parameter. We fit a logistic regression model `housing` as the predictor variable to predict if credit was approved as given by the `class` variable.

```
> model.lr.housing = glm(class ~ housing, train,
                         family='binomial')
> model.lr.housing

Call:  glm(formula = class ~ housing, family
                    = 'binomial', data = train)

Coefficients:
(Intercept)   housingown   housingrent
    0.51726      0.51540      -0.07963

Degrees of Freedom: 749 Total (i.e. Null);   747 Residual
```

```
Null Deviance:       912.9
Residual Deviance: 902.2     AIC: 908.2
```

As `housing` is a categorical variable, the model has an intercept and coefficients for the individual values of the variable except one. In the above model, the score for a data point with `housing = own` is 0.51540 plus the intercept 0.51726, whereas the score of a data point with `housing = for free` is only the intercept. We see that the coefficient for `housing = own` is greater than that for the other values of `housing`, which is consistent with our intuition.

Similar to an `lm` model, we use the `summary()` function to obtain further information about the logistic regression model including the significance values for the coefficients.

```
> summary(model.lr.housing)

Call:
glm(formula = class ~ housing, family
             = 'binomial', data = train)

Deviance Residuals:
    Min       1Q    Median       3Q       Max
-1.6354  -1.3680   0.7805   0.7805   0.9981

Coefficients:
            Estimate Std. Error z value Pr(>|z|)
(Intercept)  0.51726    0.22691   2.280   0.0226 *
housingown   0.51540    0.24720   2.085   0.0371 *
housingrent -0.07963    0.28940  -0.275   0.7832
---
Signif. codes:  0 *** 0.001 ** 0.01 * 0.05 . 0.1   1

(Dispersion parameter for binomial family taken to be 1)

    Null deviance: 912.88  on 749  degrees of freedom
Residual deviance: 902.16  on 747  degrees of freedom
AIC: 908.16

Number of Fisher Scoring iterations: 4
```

Similar to other models, we use the `predict()` function to obtain the predicted labels.

```
> predict(model.lr.housing,test)
      751       752       753       754       755       756
1.0326543 1.0326543 0.4376222 1.0326543 0.5172565 0.4376222
      757       758       759       760       761       762
1.0326543 0.5172565 1.0326543 1.0326543 1.0326543 0.4376222
...
```

The output of `predict()` contains of numeric scores for each test data point given by the model. These scores are simply the inner product between the coefficient and the data point vector: $w_0 + w_1 x$. We can apply the sigmoid function to obtain the probability of the data point x having the class $y = 1$. By default, `glm()` alphabetically assigns 0 and 1 to each value of the class variable, so in our case, we have bad $= 0$ and good $= 1$.

Calling `predict()` with the parameter `type=''response''` also outputs these probabilities.

```
> predict(model.lr.housing,test,type='response')
      751       752       753       754       755       756
0.7374302 0.7374302 0.6076923 0.7374302 0.6265060 0.6076923
      757       758       759       760       761       762
0.7374302 0.6265060 0.7374302 0.7374302 0.7374302 0.6076923
...
```

We need to apply a threshold to convert these probabilities into class labels. For example, we choose an even split of 0.5: the data points probabilities less than 0.5 will be labeled as class 0 and the ones with probabilities from 0.5 to 1 will be labeled as class 1. We use the `ifelse()` function to apply the threshold.

```
> p = predict(model.lr.housing,test,type='response')
> labels = ifelse(p > 0.5,'good','bad')
> labels
   751    752    753    754    755    756    757    758
"good" "good" "good" "good" "good" "good" "good" "good"
   759    760    761    762    763    764    765    766
"good" "good" "good" "good" "good" "good" "good" "good"
...
```

We compute the accuracy of the classifier by comparing its predicted labels against the labels of the test data.

```
> table(labels==test$class)/length(test$class)

FALSE  TRUE
0.308 0.692
```

The accuracy of our logistic regression classifier trained using the `housing` variable alone is 69.2 %.

We train a logistic regression classifier to predict the `class` variable using all features.

```
> model.lr = glm(class ~ .,train,family='binomial')
> model.lr

Call:  glm(formula = class ~ ., family
                  = 'binomial', data = train)

Coefficients:
                                    (Intercept)
                                      1.6808160
                                over.draft>=200
                                      0.9096143
                             over.draft0<=X<200
                                      0.2015547
                          over.draftno checking
                                      1.6610741
                                            ...
                                     housingown
                                     -0.5114075
```

```
                        housingrent
                         -0.9146483
                   existing.credits
                         -0.3923984
                         jobskilled
                          0.2306593
     jobunemp/unskilled non res
                          0.6427119
          jobunskilled resident
                          0.2254662
                     num.dependents
                         -0.4522923
                  own.telephoneyes
                          0.3462932
                 foreign.workeryes
                         -1.4077838

Degrees of Freedom: 749 Total (i.e. Null);  701 Residual
Null Deviance:      912.9
Residual Deviance: 657.7    AIC: 755.7
```

We obtain coefficients for all the features including each value of a categorical feature. We compute the accuracy of this classifier in a similar way.

```
> p = predict(model.lr,test,type='response')
> labels = ifelse(p>0.5,'good','bad')
> table(labels==test$class)/length(test$class)

FALSE  TRUE
0.224 0.776
```

The accuracy of our logistic regression classifier trained using all features is 77.6 %, so it performs better than using `housing` alone.

7.2.3 Support Vector Machines

Support vector machines (SVMs) is a family of binary classification algorithms. Similar to logistic regression, SVM is also a linear classifier, i.e., the classification rule is a linear function of the features. Unlike logistic regression, however, SVM is not probabilistic; the scores generated by the classifier do not have a probabilistic interpretation of the data point having a certain class label.

Each classification algorithm has a different intuitions behind it. The intuition behind SVMs is best understood visually. In Fig. 7.2, we have a two-dimensional dataset with the data points belonging to two classes: red and blue. A classification boundary assigns different labels for data points on its two sides. For now, we assume that the classes are linearly separable.

There are many potential classifiers that can separate the data points of the two classes. Figure 7.2 shows three classifiers: A, B, and C, all of which have 100 %

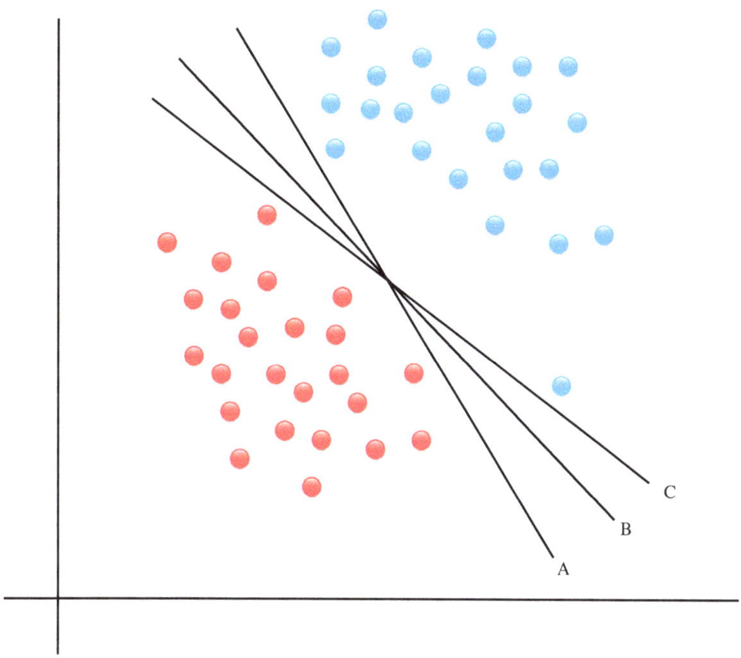

Fig. 7.2 Classifiying a two-dimensional dataset

accuracy, as they are able to classify the training dataset without any error. However, as we discussed above, accuracy of a classifier for the training dataset alone is of secondary importance; what matters is if the classifier generalizes well to new data which can be somewhat different from the training data. Ideally, a classifier should be stable or robust so that it is not affected by minor perturbations to the data.

The distance from the points of the two classes that are closest to each other is called the margin. The main idea behind SVM is to find the classifier that maximizes the margin. In turn, this classifier is equidistant from the two classes, effectively splitting the margin in half. The points that are lying on the margin are called support vectors, because they alone support or hold the classifier; by using different data points, we would end up with a different classifier.

The classifier that maximizes the margin has many useful properties. There are theoretical results that show that such a classifier does in fact generalize better than other classifiers [2]. Intuitively, we can see that the SVM classifier depends solely on the support vectors. If we perturb the dataset a little bit, the SVM classifier is less likely to change compared to a classifier that depends on all data points.

Finding such a max-margin classifier involves solving a quadratic optimization problem. We do not need to implement this ourselves as there are many software implementations of SVM available in R.[5] We use the `svm()` function of the `e1071` package, which is an interface to LIBSVM [1].

[5] For a comparison of SVM packages in R, see [5].

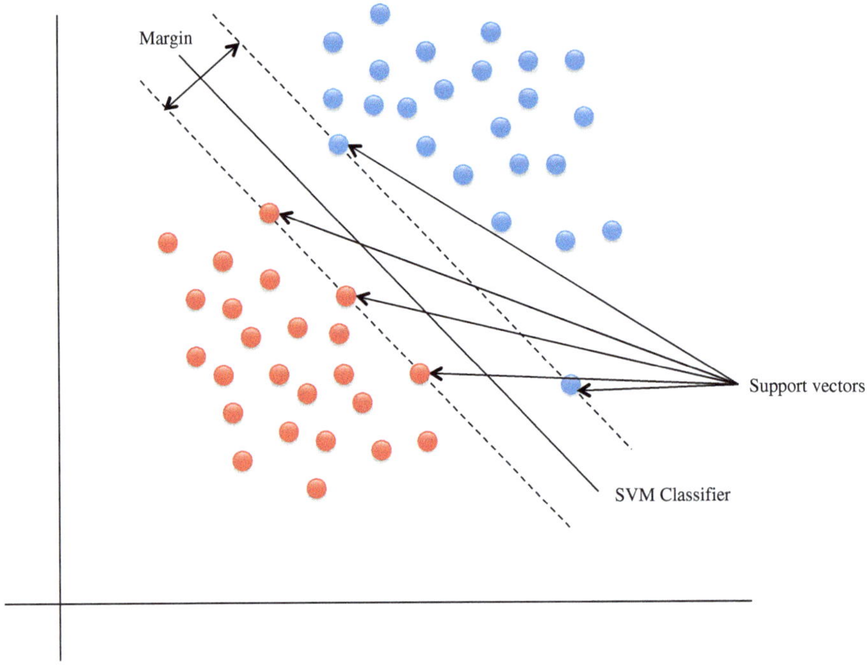

Fig. 7.3 An SVM classifier

We train an SVM classifier to predict the `class` variable below. The `svm()` function uses the same formula syntax as other classification models. We call `svm()` with the `kernel` parameter which we will look at in detail at the end of this section.

```
> model.svm = svm(class ~ ., train, kernel='linear')
> model.svm

Call:
svm(formula = class ~ ., data = train, kernel = "linear")


Parameters:
   SVM-Type:  C-classification
 SVM-Kernel:  linear
       cost:  1
      gamma:  0.01960784

Number of Support Vectors:  395
```

From the debug information, we see that the SVM classifier has 395 support vectors. We use this classifier to predict labels of the test dataset using the `predict()` function.

```
> predict(model.svm,test)
 751  752  753  754  755  756  757  758  759  760
```

```
good  bad good good good  bad good good good good
 761  762  763  764  765  766  767  768  769  770
good  bad good good good good  bad good good good
...
```

We compute the accuracy of the classifier by comparing the predicted labels with the actual labels of the test dataset.

```
> table(test$class==predict(model.svm,test))/length
                                    (test$class)

FALSE  TRUE
0.224 0.776
```

Our SVM classifier has 77.6 % accuracy.

In many practical settings, data is not quite as linearly separable as our example in Fig. 7.2. For nonlinearly separable data, we train an SVM classifier by allowing for a certain amount of misclassification which is also called *slack*. We can control the slack using the `cost` parameter of `svm()`. By default `cost` is set to 1, we can sometimes obtain a more accurate classifier by setting the appropriate value for this parameter.

In the following example, we increase the accuracy of the classifier to 78 % by setting `cost=0.25`.

```
> model.svm.cost = svm(class ~ .,train,
                       kernel='linear',cost=0.25)
> table(test$class==predict(model.svm.cost,test))/
                        length(test$class)

FALSE  TRUE
 0.22  0.78
```

7.2.3.1 Kernel Trick

An alternative strategy of dealing with nonlinearly separable data is to use the kernel trick.[6] The idea behind the kernel trick is to project the data points of the training data into a higher dimensional space, in which the dataset is linearly separable. This projection could even be into an infinite-dimensional space.

SVMs have the advantage that it is very efficient to train the classifier without actually making projections of the data points; storing infinite-dimensional data

[6] Kernels unfortunately mean two different things in statistics. The kernel functions that we refer to here have some similarities to the ones we saw in Sect. 6.3.2, but there are some important theoretical differences as well.

points would be infeasible. Instead, we only consider the relative distances between the data points in the projected space, as given by the kernel function.

Some of the commonly used kernel functions include polynomial, sigmoid, and perhaps the most popular: Gaussian or radial kernel. In the `svm()` function, we can specify the type of the kernel using the `kernel` argument. We train an SVM classifier with radial kernel below.

```
> model.svm.radial = svm(class ~ ., train, kernel='radial')
> model.svm.radial

Call:
svm(formula = class ~ ., data = train, kernel = "radial")


Parameters:
   SVM-Type:  C-classification
 SVM-Kernel:  radial
       cost:  1
      gamma:  0.01960784

Number of Support Vectors:  470
```

Note that the number of support vectors has increased when using the radial kernel.

For our dataset, using the radial kernel leads to a small decrease in accuracy: we see that it goes down to 74.4 %.

```
> table(test$class==predict(model.svm.radial,test))/
                        length(test$class)

FALSE  TRUE
0.256 0.744
```

Note that we have used default values for the kernel parameters. Obtaining high classification accuracy with kernels usually involves tuning these parameter values either through a grid search. Increasingly, there have been techniques proposed to learn these parameters directly from the data as well.

7.3 Nonparametric Classification Models

Similar to the nonparametric models for regression that we saw in Sect. 6.3, we can also use nonparametric models for classification. In nonparametric models, there are no explicit modeling assumptions related to the shape or form of the classifier. In this section, we look at two important nonparametric classifiers: nearest neighbors and decision trees.

7.3.1 *Nearest Neighbors*

Nearest neighbors (NN) are perhaps the simplest to understand, yet one of the most powerful nonparametric classification models. To classify a test data point using an NN model, we use the training data points in its neighborhood. The predicted class label of the test data point is given by the class label of the majority of the training data points in its neighborhood. Here, neighborhood refers to the group of data points in geometric proximity to each other as defined by a distance metric. In NN classifiers, there is no separate model fitting or training step; we are using the training dataset itself as the model. We obtain the predicted value of a test data point only by looking up its neighbors in the training dataset.

The power of NN classifiers comes from the fact that the classifier we learn is relying entirely on the geometry of the training data without assuming any specific distribution. These classifiers work well for nonlinear datasets and even for datasets that are not linearly separable. Although NN classifiers work quite accurately with low dimensional data; there are often issues when working with sparse datasets where we have large number of dimensions and fewer data points. In such cases, there are only a handful of data points from the training dataset in the neighborhood of a test data point. On the other hand, NN classifiers work well when we have a large number of data points. The only parameter we use for NN is the number of data points to consider in the neighborhood, usually denoted by k. In practice, we choose k based on the value that gives the highest accuracy on test dataset or better yet, using crossvalidation.

There are many distance metrics used with NN classifiers: the most popular one is the Euclidean distance, which is given by the linear distance between two data points. Given two d-dimensional data points x_1 and x_2, the Euclidean distance between them is given by:

$$\text{Euclidean}(x_1, x_2) = \sqrt{\sum_i (x_{1i} - x_{2i})^2}.$$

As Euclidean distance uses the square of the difference between the data point components, it is also referred to as the l_2 distance.

Another popular metric is the city-block or Manhattan distance, which is the distance traveled along a grid to reach one point from the next. For the data points x_1 and x_2, the Manhattan distance between them is given by:

$$\text{Manhattan}(x_1, x_2) = \sum_i |x_{1i} - x_{2i}|.$$

As the Manhattan distance uses the absolute value of the difference between the data point components, it is also called the l_1 distance. A generalization of the Euclidean and Manhattan distance metrics is the Minkowski distance. For the data points x_1

and x_2, the Minkowski distance of order p between them is given by:

$$\text{Minkowski}(x_1, x_2) = \left(\sum_i |x_{1i} - x_{2i}|^p \right)^{\frac{1}{p}}.$$

Setting $p = 2$ gives us Euclidean distance, while setting $p = 1$ gives us Manhattan distance.

All data points in the neighborhood of a data point are not equally important; in a group of k data points, some of the data points could be closer or further away from the test data point. A variant of NN classifiers is weighted nearest neighbor classifiers, where we assign weights to data points in the neighborhood based on the distance from the test data point. Here, the class label of the closest neighbors of the test data point will affect the predicted class label more than the labels of the distant neighbors. We can use a kernel function to assign weight based on the distance metric; here the weight should be maximum for zero distance and it should decrease as the distance increases. Conceptually, weighted NN is similar to kernel regression that we looked at in Sec. 6.3.2; we can reuse the same kernel functions here as well.

There are many R packages for NN classifiers: some of the popular implementations include `kknn`, `FNN`, `knn` function of the `class` package, `IBk` function of the `RWeka` package. We use the `kknn` package here. We use the eponymous `kknn()` function to obtain a weighted NN classifier. We need to provide the formula denoting the class label, and the data frames for training and test dataset.

The `kknn()` function uses Minkowski distance as its metric. We can specify the order p using the `distance` parameter which is by default set to 2, corresponding to Euclidean distance. We first use unweighted NN algorithm; this is equivalent to setting the `kernel` parameter to `'rectangular.'` By default, `kknn()` sets the value of `k` to 7.

```
> library(kknn)
> model=kknn(class ~ ., train,test,kernel='rectangular')
> model

Call:
kknn(formula = class ~ ., train = train, test = test,
     kernel = "rectangular")

Response: "nominal"
```

Printing the `model` variable does not display too much information. As with other models, we obtain the predicted class labels using the `predict()` function.

```
> table(predict(model,test) == test$class)/length(test$class)

FALSE  TRUE
0.236 0.764
```

The accuracy of unweighted NN with `k` = 7 is 76.4 %. Sometimes we can obtain better accuracy by trying out different values of `k`.

```
> model.5 = kknn(class ~ ., train,
```

```
                test,kernel='rectangular',k=5)
> table(predict(model.5,test) == test$class)/length
                                   (test$class)

FALSE  TRUE
0.228 0.772
```

In this case, the model achieves 77.2 % accuracy using k = 5. Similarly, we can also experiment with other kernels. The model achieves a higher accuracy of 78.8 % using Gaussian kernel with k = 8.

```
> model.gaussian = kknn(class ~ ., train,test,
                        kernel='gaussian',k=8)
> table(predict(model.gaussian,test) == test$class)/
                 length(test$class)

FALSE  TRUE
0.212 0.788
```

7.3.2 Decision Trees

As we mentioned in Sect. 6.3.3, we can use decision trees as much for classification as well as regression. Decision trees work exactly the same way for predicting categorical labels instead of a numeric target variable.

We also use the `rpart` package to fit decision trees. We use the training data to fit a decision tree to predict the `class` variable below.

```
> model.dt = rpart(class ~ .,train)
```

We plot the tree in Fig. 7.4 using the following code.

```
> plot(model.dt,uniform=T)
> text(model.dt,pretty=T,minlength=20)
```

We use `predict()` to obtain the predictions using this model. By default, the predict function returns the class probabilities. We obtain the class labels by calling `predict()` with the argument `type='class.'`

```
> predict(model.dt,test,type='class')
 751  752  753  754  755  756  757  758  759  760
good good good good good  bad good good good good
 761  762  763  764  765  766  767  768  769  770
good good good good good good good good good good
...
```

We compare the predicted classes to obtain the accuracy of this classifier.

```
> p=predict(model.dt,test,type='class')
```

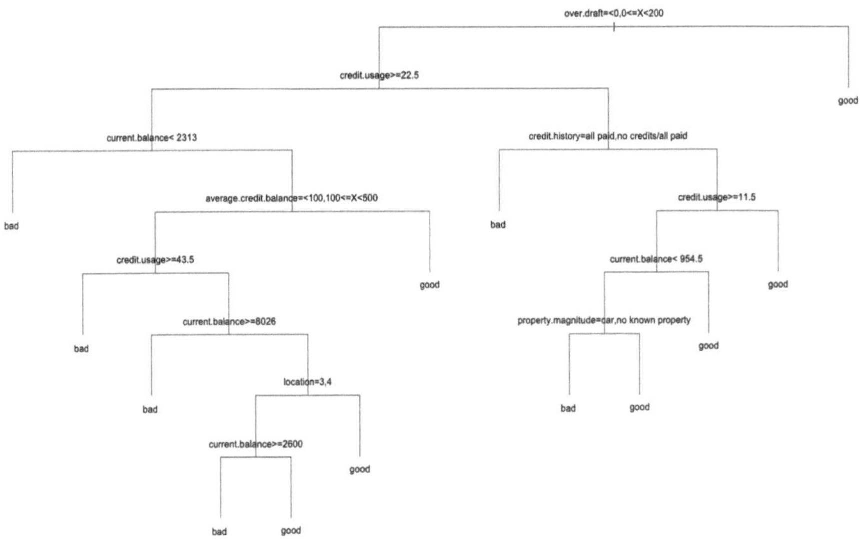

Fig. 7.4 Decision tree classifier

```
> table(test$class == p)/length(test$class)

FALSE  TRUE
0.252 0.748
```

The accuracy of our decision tree classifier is 74.8 %.

It is also useful to prune decision trees for classification. There are a lot of branches in the tree in Fig. 7.4; we set the `cp` parameter to prune some of the branches off by setting it to 0.04.

```
> model.dt.prune = rpart(class ~ .,train,cp=0.04)
```

This results in a much simpler tree which we plot in Fig. 7.5.

Apart from simplicity, using a pruned tree has accuracy benefits as well. We see that the decision tree in `model.dt.prune` has an accuracy of 75.6 % which is better than less pruned counterpart.

```
> p=predict(model.dt.prune,test,type='class')
> table(test$class == p)/length(test$class)

FALSE  TRUE
0.244 0.756
```

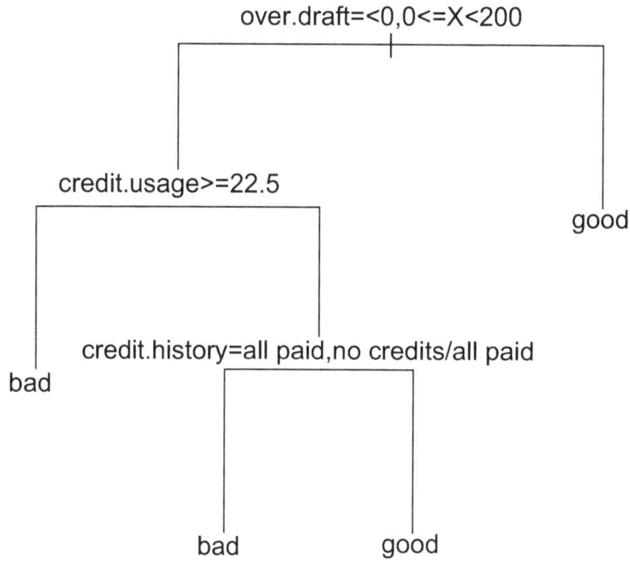

Fig. 7.5 Pruned decision tree classifier

7.4 Chapter Summary

In this chapter, we looked at classification where we fit a model to classify data points into multiple classes. In classification models, the target variable or class label is categorical, whereas in regression, the target variable is numeric. Similar to regression, classification models are of two types: parametric and nonparametric.

In parametric models, we first looked at naive Bayes (NB) model which classifies the data points using the data likelihood probabilities. This model makes the assumption of independence of features. We use the naiveBayes() function of the e1071 package to fit a naive Bayes model. In logistic regression (LR), we fit the class probabilities using a sigmoid function and obtain the coefficients of this model using gradient descent. We use the glm() function from the R base package to train a logistic regression classifier. A support vector machine (SVM) classifier learns the hyperplane that separates the data points between the two classes with the maximum margin, which is the closest distance between the data points of the two classes. Another benefit of SVM is the kernel trick, where we can classify a nonlinearly separable dataset by efficiently projecting the data points into a higher dimensional space. We use the svm() function of the e1071 package to train an SVM classifier.

In nonparametric models, we looked at nearest neighbor (NN) models for classification. In these models, we do not learn any model parameters, but make predictions based on the data points from the training data that are in the neighborhood of the given test data point. We used the `kknn` package for obtaining NN classifiers. We also looked at decision tree based models for classification. Similar to the decision tree models for regression, we split the dataset at every node based on the value of a variable and classify the data points in a leaf node based on the aggregate values of the class labels. We used the rpart package to fit the decision trees for classification using the classification and regression tree (CART) algorithm.

References

1. Chang, C.-C., & Lin, C.-J. (2011). LIBSVM: A library for support vector machines. ACM Transactions on Intelligent Systems and Technology, *2:27,* 1–27:27. http://www.csie.ntu.edu.tw/cjlin/libsvm. Accessed 1 Aug 2014.
2. Cortes, C., & Vapnik, V. (1995). Support-vector networks. *Machine Learning, 20*(3), 273–297.
3. Geisser, S. (1993). *Predictive inference*. UK: Chapman and Hall.
4. Graham, P. (2002). A plan for spam. http://www.paulgraham.com/spam.html. Accessed 1 Aug 2014.
5. Karatzoglou, A., & Meyer, D. (2006). Support vector machines in R. *Journal of Statistical Software, 15,* 1–28.

Chapter 8
Text Mining

8.1 Introduction

In this chapter we consider the problem of analyzing text data. Text is perhaps the most ubiquitous form of data. The amount of text data available is staggering: from books or articles, to web pages, to social media. All of these form rich sources of data which we can exploit to gain valuable information. Analysis of text data has many practical applications including information retrieval, social network analysis, spam filtering, and sentiment analysis.

Text mining, or text analytics as it is alternatively known, is the task of extracting information from text data. In the previous chapters, we looked at using R for performing data analysis in various settings. There was a common element in all of the settings: the input data that we used was almost always *structured*, i.e., consisting of data points for a given set of variables either numeric or categorical. Text data is *unstructured*. Although text written in any human language is governed by the rules of grammar, it lacks the clearly defined variables that we have when analyzing structured data. Apart from linguistic rules, text carries information, such as describing an object or presenting a point of view. In turn, the goal of text mining is to extract this latent information from text data as objectively as possible.

Text mining is an umbrella term for different types of analysis that we perform over text. The analysis tasks range from simple word-frequency based analysis to more complex tasks such as classification. In fact, we can easily adapt classical data analysis techniques such as exploratory data analysis and classification when working with text data. We only need to use the proper representation of text so that the unstructured text data can be encoded as structured or semi-structured data. In this chapter we will look at various data structures to represent text documents.

The text mining functionality in R is built around the `tm` package which we review in this chapter.[1] The `tm` package has a well-designed data structures and utility

[1] Different versions of the tm package have slight differences in function names. In this chapter we use tm package version 0.5–10.

M. A. Pathak, *Beginning Data Science with R*,
DOI 10.1007/978-3-319-12066-9_8

functions for representing text data. Another benefit of `tm` is that it is convenient to export the data so that it can be used in other R functions and packages.

8.2 Dataset

In this chapter, we look at a collection of reviews for 2000 movies.[2] The reviews are classified into positive and negative according to their overall sentiment. There are equal number of positive and negative reviews. Due to the sentiment labeling, this dataset leads us to various interesting analysis tasks including identifying the most positive and negative words, and classifying if a given document has a positive or negative sentiment.

The movie review dataset is located in the directory `review_polarity`. All of the movie reviews are in the form of flat text files. The positive movie reviews are located in the directory `review_polarity/txt_sentoken/pos/`, while the negative movie reviews are located in the directory `review_polarity/txt_sentoken/neg/`.

Before we begin our analysis, we load the `tm` package using the library function.

```
> library(tm)
```

8.3 Reading Text Input Data

The fundamental data structure of tm is the TextDocument object, which, as the name suggests, represents a text document. After that, we have the Corpus object which represents a collection of text documents.[3] Finally, we have the TextRepository object which is a collection of multiple Corpus objects.

Text documents are available in heterogeneous data formats: from ubiquitous plain text, PDF, HTML formats to more special purpose formats such as RCV1. Plain text is the easiest format to process as it contains text alone without additional metadata. In many other text processing packages, we first need to use external tools to convert data into plain text. One of the strengths of the tm package is its support to read data from various file formats. This functionality is available through the reader functions for each file format. We can obtain the list of reader functions using getReaders().

---

[2] We use the polarity dataset v2.0 available for download at http://www.cs.cornell.edu/people/pabo/movie-review-data/.

[3] In linguistics, corpus, which literally means a body, is the term used to refer to a collection of documents.

```
> getReaders()
[1] "readDOC"
[2] "readPDF"
[3] "readReut21578XML"
[4] "readReut21578XMLasPlain"
[5] "readPlain"
[6] "readRCV1"
[7] "readRCV1asPlain"
[8] "readTabular"
[9] "readXML"
```

Apart from file formats, text documents are available from different sources: text files, web pages with a URL, or even in existing R data frame or vector objects. The data frames are useful for intermediate data storage when we are using an external source such as an SQL database. In tm, the data source is encapsulated as the Source object. We have different types of Source objects for each specialized data source. We can see the list of the available Source objects using getSources().

```
> getSources()
[1] "DataframeSource" "DirSource"
[3] "ReutersSource"   "URISource"
[5] "VectorSource"
```

We use DirSource when we have a directory containing text documents as the case is with our dataset.

To create a Corpus object, we require a Source object and a reader function. If we do not specify a reader function, readPlain() is used by default. We create a corpus of positive documents from our dataset directory `review_polarity/txt_sentoken/pos`. As we discussed above, this directory contains 1000 text files.

```
> source.pos = DirSource('review_polarity/txt_
               sentoken/pos')
> corpus.pos = Corpus(source.pos)
```

Similarly, we create the corpus of negative documents from the directory `'review_polarity/txt_sentoken/neg`.

```
> source.neg = DirSource('review_polarity/txt_
               sentoken/neg')
> corpus.neg = Corpus(source.neg)
```

The corpus object `corpus.pos` as well as `corpus.neg` contain the 1000 text documents. Printing the corpus.pos object confirms this with a human-readable message.

```
> corpus.pos
A corpus with 1000 text documents
```

The corpus object is indexable as a vector, so we can use the length() function to determine its length.

```
> length(corpus.pos)
[1] 1000
```

Similarly, we use vector indexing to obtain individual text documents in the corpus object.

```
> corpus.pos[1]
A corpus with 1 text document
```

Printing the text document `corpus.pos[1]` on the console is equivalent to using the show() function.

```
> show(corpus.pos[1])
A corpus with 1 text document
```

The inspect() function allows us to view the contents of the text document along with its metadata. As we have not yet assigned the metadata to this document, we can only see the text contents after the file name `$cv000_29590.txt`. This is the same text that we can see if we opened the file `review_polarity/txt_sentoken/pos/cv000_29590.txt` in a text editor.

```
> inspect(corpus.pos[1])
A corpus with 1 text document

The metadata consists of 2 tag-value pairs and a data frame
Available tags are:
  create_date creator
Available variables in the data frame are:
  MetaID

$cv000_29590.txt
films adapted from comic books have had plenty of success,
whether they're about superheroes (batman, superman, spawn), or
geared toward kids (casper) or the arthouse crowd (ghost world),
but there's never really been a comic book like from hell before.
for starters, it was created by alan moore (and eddie campbell),
who brought the medium to a whole new level in the mid '80s with a
12-part series called the watchmen.
to say moore and campbell thoroughly researched the subject of jack
the ripper would be like saying michael jackson is starting to look a
little odd.
...
```

8.4 Common Text Preprocessing Tasks

Once we have our text documents loaded into R, there are a few preprocessing tasks that we need to do before we can begin our analysis. In most cases, these tasks are a set of transformations that we apply on the text document. For example, when we are trying to find the frequencies of word occurences in a corpus, it makes sense to remove the punctuation symbols from the text, otherwise commonly occurring punctuation symbols such as full-stop and comma would also start appearing in our word counts.

The tm package has the `tm_map()` function that applies a transformation to a corpus or a text document and also takes a transformation function as input. To remove the punctuation symbols, we use the `removePunctuation()` function.

```
> doc = tm_map(corpus.pos[1],removePunctuation)
> inspect(doc)
A corpus with 1 text document

The metadata consists of 2 tag-value pairs and a data frame
Available tags are:
  create_date creator
Available variables in the data frame are:
  MetaID

$cv000_29590.txt
films adapted from comic books have had plenty of success whether
theyre about superheroes batman superman spawn or geared toward
kids casper or the arthouse crowd ghost world but theres never
really been a comic book like from hell before
for starters it was created by alan moore and eddie campbell who
brought the medium to a whole new level in the mid 80s with a 12part
series called the watchmen
to say moore and campbell thoroughly researched the subject of jack
the ripper would be like saying michael jackson is starting to look a
little odd
...
```

As compared to the output of `inspect(corpus.pos[1])`, all of the punctuation symbols are now removed. The `tm_map()` function does not make any changes to the input document but returns another text document with the modified text.

The `tm_map()` function supports many other transformations: we can see the list of available transformations using the getTransformations() function.

```
> getTransformations()
[1] "as.PlainTextDocument"
[2] "removeNumbers"
[3] "removePunctuation"
[4] "removeWords"
```

```
[5] "stemDocument"
[6] "stripWhitespace"
```

Two other important transformations are stop word removal using the removeWords() transformation and stemming using the stemDocument() transformation.

8.4.1 Stop Word Removal

All words in a text document are not equally important: while some words carry meaning, others play only a supporting role. The words in the latter category are called stop words. The list of stop words is of course specific to a language. Examples of stop words in English include "the," "is," "a," "an." As these words occur very frequently in all documents, it is better to first remove them from the text.

There are lists of stop words compiled for most languages. The `tm` package makes such lists available for English and a few other European languages through the stopwords() function. We need to specify the language name as an argument to this function, but it returns the list for English by default.

```
> stopwords()
  [1] "i"          "me"
  [3] "my"         "myself"
  [5] "we"         "our"
  [7] "ours"       "ourselves"
  [9] "you"        "your"
 [11] "yours"      "yourself"
 [13] "yourselves" "he"
 [15] "him"        "his"
 [17] "himself"    "she"
 [19] "her"        "hers"
 [21] "herself"    "it"
...
```

To remove the stop words from a text document, we use the removeWords() transformation in the `tm_map()` function. We also need to provide the list of stopwords as the third argument.

```
> doc = tm_map(doc,removeWords,stopwords())
> inspect(doc)
A corpus with 1 text document
...
$cv000_29590.txt
films adapted comic books plenty success whether theyre
superheroes batman superman spawn geared toward kids casper
arthouse crowd ghost world theres never really comic book like
hell starters created alan moore eddie campbell brought medium
 whole new level mid 80s 12part series called watchmen
 say moore campbell thoroughly researched subject jack ripper
```

```
 like saying michael jackson starting look little odd
 book graphic novel will 500 pages long includes nearly 30
 consist nothing footnotes words dont dismiss film source
...
```

As we can see, the words such as "from," "have," "had" have been removed from the text.

In the example above, we used the generic list of English stop words. In many scenarios, this is not ideal as we have *domain-specific* stop words. In a corpus of movie reviews, words such as "movie" or "film" occur fairly often, but are devoid of useful content, as most of the documents would contain them. We therefore need to do a frequency analysis to identify such commonly occuring words in a corpus. We shall see an example of frequency analysis further in the chapter.

It is straightforward to remove our own list of stopwords in addition to the generic list available in the stopwords() function. We simply concatenate the two lists using the c() function.

```
> doc = tm_map(doc,removeWords,
              c(stopwords(),'film','films','movie','movies'))
> inspect(doc)
A corpus with 1 text document
...
$cv000_29590.txt
 adapted comic books plenty success whether theyre superheroes
batman superman spawn geared toward kids casper arthouse
crowd ghost world theres ever really comic book like hell
 starters created alan moore eddie campbell brought medium
whole new level mid 80s 12part series called watchmen
 say moore campbell thoroughly researched subject jack ripper
like saying michael jackson starting look little odd
...
```

8.4.2 Stemming

In the above example of stopword removal, we need to specify multiple forms of the same word such as "film" and "films." As an alternative to this, we can use a stemming transformation on the document. In linguistics, stemming is the task of replacing each word in a document to its stem or basic form, e.g., "films" by "film." Note that multiple words typically stem to the same word, e.g., the words "filmed," "filming," and "filmings" all would stem to the same word "film."

In most cases, the words and their stemmed counterpart carry a similar meaning or at least share something in common; in the above example, they have something to do with a "film." Stemming greatly reduces the redundancy in text. This finds applications in many NLP tasks: in the case of word frequency analysis over a stemmed corpus, the counts of individual words get added up in the counts for their stemmed form. If the word "film" occurs 100 times in the original corpus, and "films" occurs 50 times, in the stemmed corpus, the count for the word "films" will be 150, which is a more representative count. Stemming plays an important role in information retrieval: most search engines perform stemming on the search queries and text documents so that we can find documents containing different forms of the same words that occur in the query. A search using the word "cats" should also match a document with the word "cat."

Automatic stemming algorithms or stemmers have more than a 50 year history. The most popular stemmer for English is the Porter stemmer. There are software implementations of the Porter stemmer for most programming environments. In the tm package, the Porter stemmer is available through the stemDocument() transformation.

We use the stemDocument() function as an argument to `tm_map()`.

```
> doc = tm_map(doc,stemDocument)
> inspect(doc)
A corpus with 1 text document
...
$cv000_29590.txt
adapt comic book plenti success whether theyr superhero batman
superman spawn gear toward kid casper arthous crowd ghost
world there never realli comic book like hell
starter creat alan moor eddi campbel brought medium whole
new level mid 80s 12part seri call watchmen
say moor campbel thorough research subject jack ripper like say
michael jackson start look littl odd
...
```

As we can see words are replaced by their stemmed versions, e.g., "adapted" by "adapt" and "books" by "book."

8.5 Term Document Matrix

The term document matrix is a fundamental data structure in text mining. In natural language processing, the term document matrix is also referred to as *bag of words*. Term document matrix is the data representation where the text data is stored with some form of structure, making it the bridge between unstructured data and structured data. Most of the text mining algorithms involve performing some computation on this representation.

As the name suggests, the term document matrix is a matrix containing individual words or terms as rows and documents as columns. The number of rows in the matrix is the number of unique words in the corpus, and the number of columns is the number of documents. The entry in the matrix is determined by the weighting function. By default, we use the term frequency, that is the count of how many times the respective word occurs in the document. If a document A contains the text "this is a cat, that is a cat.", the column for document A in the term document matrix will contain entries with value 1 each for "this" and "that," and entries with value 2 for "is," "a," and "cat." In this way, the term document matrix only contains the word occurence counts and ignores the order in which these words occur. An alternative formulation is the document term matrix, with the documents as rows and individual terms as columns. In this section, we only consider term document matrices.

In tm, the term document matrix is represented by the TermDocumentMatrix object. We create a term document matrix for either a document or a corpus using its constructor function TermDocumentMatrix(). We create the term document matrix for the corpus of positive sentiment documents `corpus.pos` below.

```
> tdm.pos = TermDocumentMatrix(corpus.pos)
> tdm.pos
A term-document matrix (36469 terms, 1000 documents)

Non-/sparse entries: 330003/36138997
Sparsity           : 99%
Maximal term length: 79
Weighting          : term frequency (tf)
```

As with TextDocument objects, printing a TermDocumentMatrix object displays summary information. There are 36,469 terms in corpus with 1000 documents. The nonsparse entries refer to the number of entries in the matrix with nonzero values. Sparsity is the percentage of those entries over all the entries in the matrix. As we can see the sparsity is 99 %. This is a common phenomenon with text corpora as most of them are very sparse. Most documents of our corpus contain only a few hundred terms, so the remaining 36,000 or so entries for each document are zero. Maximal term length is the length of the longest word. As we mentioned above, we are using the term frequency weighting function.

We view the contents of the TermDocumentMatrix object using the inspect() function. As displaying the TermDocumentMatrix for the entire corpus of 1000 documents takes a lot of space, we only display the TermDocumentMatrix computed over the corpus of the three documents here.

```
> tdm.pos3 = TermDocumentMatrix(corpus.pos[1:3])
> inspect(tdm.pos3)
A term-document matrix (760 terms, 3 documents)

Non-/sparse entries: 909/1371
Sparsity           : 60%
Maximal term length: 16
Weighting          : term frequency (tf)

               Docs
Terms           cv000_29590.txt cv001_18431.txt cv002_15918.txt
  _election                   0               2               0
  _election_                  0               7               0
  _ferris                     0               1               0
  _rushmore_                  0               6               0
  '80s                        1               0               0
  102                         1               0               0
  12-part                     1               0               0
```

```
  1888                        1               0               0
  500                         1               0               0
  abberline                   2               0               0
  ably                        1               0               0
  about                       4               0               2
  absinthe                    1               0               0
  absolutely                  0               0               1
  accent                      2               0               0
  across                      0               0               1
  acting                      1               0               1
  acts                        1               0               0
  actually                    1               0               1
  adapted                     1               0               0
  add                         0               1               0
  adding                      0               0               1
  ...
```

We see that the stop words are present in the term document matrix. There are two ways of removing stopwords and performing additional transformations like stemming and punctuation removal. We can either apply these transformations over the corpus object using `tm_map()` before calling `TermDocumentMatrix()`. The second option is to pass a list of transformation in the `control` parameter of `TermDocumentMatrix()`. We create a term document matrix from the first three documents of `corpus.pos` after removing the stop words, punctuations, and applying stemming.

```
> tdm.pos3 = TermDocumentMatrix(corpus.pos[1:3],
               control=list(stopwords = T,
               removePunctuation = T, stemming = T))

> inspect(tdm.pos3)

A term-document matrix (613 terms, 3 documents)

Non-/sparse entries: 701/1138
Sparsity           : 62%
Maximal term length: 14
Weighting          : term frequency (tf)


             Docs
Terms          cv000_29590.txt cv001_18431.txt cv002_15918.txt
  102                        1               0               0
  12part                     1               0               0
  1888                       1               0               0
  500                        1               0               0
  80s                        1               0               0
  abberlin                   2               0               0
  abli                       1               0               0
  absinth                    1               0               0
  absolut                    0               0               1
  accent                     2               0               0
  across                     0               0               1
```

```
act                    2                0                1
actual                 1                0                1
adapt                  1                0                0
add                    0                1                0
...
```

As we can see, this term document matrix does not contain stopwords, punctuation words starting with _ and only contains stemmed words. We apply the same transformations to the entire corpus object below.

```
> tdm.pos = TermDocumentMatrix(corpus.pos,
                control = list(stopwords = T,
                removePunctuation = T, stemming = T))
> tdm.pos
A term-document matrix (22859 terms, 1000 documents)

Non-/sparse entries: 259812/22599188
Sparsity           : 99%
Maximal term length: 61
Weighting          : term frequency (tf)
```

With these transformations, we see that the number of terms in `tdm.pos` has decreased from 330,003 to 259,812: a 21.27 % reduction. This alone, however, does not significantly reduce the sparsity of the term document matrix.

8.5.1 TF-IDF Weighting Function

In the above term document matrices, we are using the term frequency (TF) weighting function. Another popular weighting function is term frequency-inverse document frequency (TF-IDF) which is widely used in information retrieval and text mining.

As we mentioned above, TF is the count of how many times a term occurs in a document. More often a word appears in the document, higher its TF score. The idea behind inverse document frequency (IDF) is that all words in the corpus are not equally important. IDF is a function of the fraction of documents in which the term appears in. If a corpus has N documents, and a term appears in d of them, the IDF of the term will be $\log(N/d)$.[4] A term appearing in most of the documents will have a low IDF as the ratio N/d will be close to 1. A term that appears in only a handful of documents will have a high IDF score as the N/d ratio will be large.

The TF-IDF weighting score for a term-document pair is given by its term frequency multiplied by the IDF of the term. If a term that infrequently occurs in the corpus occurs in a given document multiple times, that term will get a higher score

[4] log is used as a smoothing function. Alternative formulations of IDF use other such functions.

for the given document. Alternatively, terms that are frequently occuring in the corpus such as stop words will have a low IDF score as they appear in most documents. Even if such words have a high TF score for a document, their TF-IDF score will be relatively low. Due to this, TF-IDF weighting function has an effect similar to stop word removal.

We specify the TF-IDF weighting function `weightTfIdf()` in the control parameter of `TermDocumentMatrix()`.

```
> tdm.pos3 = TermDocumentMatrix(corpus.pos[1:3],
                   control = list(weighting = weightTfIdf,
                   stopwords = T, removePunctuation = T,
                   stemming = T))
> inspect(tdm.pos3)
A term-document matrix (613 terms, 3 documents)

Non-/sparse entries: 665/1174
Sparsity           : 64%
Maximal term length: 14
Weighting          : term frequency - inverse document
                     frequency (normalized)

              Docs
Terms           cv000_29590.txt cv001_18431.txt cv002_15918.txt
  102               0.004032983     0.000000000     0.000000000
  12part            0.004032983     0.000000000     0.000000000
  1888              0.004032983     0.000000000     0.000000000
  500               0.004032983     0.000000000     0.000000000
  80s               0.004032983     0.000000000     0.000000000
  abberlin          0.008065967     0.000000000     0.000000000
  abli              0.004032983     0.000000000     0.000000000
  absinth           0.004032983     0.000000000     0.000000000
  absolut           0.000000000     0.000000000     0.007769424
  accent            0.008065967     0.000000000     0.000000000
  across            0.000000000     0.000000000     0.007769424
  act               0.002976908     0.000000000     0.002867463
  actual            0.001488454     0.000000000     0.002867463
  adapt             0.004032983     0.000000000     0.000000000
  add               0.000000000     0.004620882     0.000000000
  ...
```

The entries in the term document matrix are all fractional due to the TF-IDF weighting function.

Some of the other commonly used weighting functions include binary weighting where we assign a score of 0 or 1 to a term-document pair depending on whether the term appears in the document while ignoring the frequency of occurence. This weighting function is available through `weightBin()`.

8.6 Text Mining Applications

With these data structures at hand, we now look at a few algorithms for text mining.

8.6.1 Frequency Analysis

The first and simplest text mining algorithm is based on counting the number of words in a corpus. Despite its simplicity, frequency analysis provides a good high-level overview of the corpus contents. The most frequently occuring words in the corpus usually indicate what the corpus is about.

We can determine the frequency of occurence for a word simply by summing up the corresponding row in the term document matrix. The `findFreqTerms()` function does a similar computation. It also takes a lower and upper threshold limits as inputs and returns the words with frequencies between those limits.

We first construct the term document matrices for the two corpus objects `corpus.pos` and `corpus.neg`.

```
> tdm.pos = TermDocumentMatrix(corpus.pos,
                    control = list(stopwords = T,
                    removePunctuation = T, stemming = T))

> tdm.neg = TermDocumentMatrix(corpus.neg,
                    control = list(stopwords = T,
                    removePunctuation = T, stemming = T))
```

We apply `findFreqTerms()` to the term document matrix of the corpus of positive reviews `tdm.pos` below. We use a lower threshold of 800 and do not specify the upper threshold. By default, the upper threshold is set to infinity.

```
> findFreqTerms(tdm.pos,800)
 [1] "also"    "best"    "can"     "charact" "come"    "end"
 [7] "even"    "film"    "first"   "get"     "good"    "just"
[13] "life"    "like"    "look"    "love"    "make"    "movi"
[19] "much"    "one"     "perform" "play"    "scene"   "see"
[25] "seem"    "stori"   "take"    "thing"   "time"    "two"
[31] "way"     "well"    "will"    "work"    "year"
```

The `findFreqTerms()` function returns a vector of words sorted in alphabetical order. For the term document matrix `tdm.pos`, the frequent words includes words with positive sentiments such as "best," "good," and "well." This list also includes filler words such as "also," "can," "even" that do not carry too much meaning by themselves. These words are present in the term document matrix even after we applied stop word removal, so it appears that these words are not present in the stop word list. Frequency analysis is a useful tool to augment the stop word list for a corpus.

Similarly, we apply `findFreqTerms()` to the term document matrix of the corpus of negative reviews `tdm.neg`.

```
> findFreqTerms(tdm.neg,800)
 [1] "bad"     "can"     "charact" "come"    "end"     "even"
 [7] "film"    "first"   "get"     "good"    "just"    "like"
[13] "look"    "make"    "movi"    "much"    "one"     "play"
[19] "plot"    "scene"   "see"     "seem"    "stori"   "take"
[25] "thing"   "time"    "two"     "way"     "will"
```

We see that the frequent terms contain words with negative sentiment such as "bad," but also contain words such as "good." This indicates that frequency analysis is not alone suitable to determine if a given review is positive or negative.

We can also use frequency analysis to identify frequently co-occuring words. Such words are refered to as word associations. The `findAssocs()` function takes a term document matrix object along with an input word and returns its word associations along with correlation scores. The correlation score is the degree of confidence in the word association. When using `findAssocs()`, we also need to specify a lower limit for the correlation score so that the function only returns the word associations with correlations greater than or equal to that limit. We find the associations for the word "star" with the correlation limit 0.5.

```
> findAssocs(tdm.pos,'star', 0.5)
                         star
trek                     0.63
enterpris                0.57
picard                   0.57
insurrect                0.56
jeanluc                  0.54
androidwishingtobehuman  0.50
anij                     0.50
bubblebath               0.50
crewmat                  0.50
dougherti                0.50
everbut                  0.50
harkonnen                0.50
homeworld                0.50
iith                     0.50
indefin                  0.50
mountaintop              0.50
plunder                  0.50
reassum                  0.50
ruafro                   0.50
soran                    0.50
unsuspens                0.50
verdant                  0.50
youthrestor              0.50
```

We see that the strongest associations for the word "star" are "trek" and "enterpris" arising from the movie Star Trek.

8.6.2 *Text Classification*

Text classification is the task of classifying text documents into two or more classes based on their content. Similar to the classification techniques we discussed in Chap. 7, we learn a classifier from a set of text documents annotated with class labels. Text classification finds application in a wide variety of settings: spam filtering, where we classify emails into spam or not spam, document categorization, where we classify documents into a prespecified taxonomy, and sentiment analysis. Our movie reviews dataset is a good setting where we can apply sentiment analysis. Using the positive and negative movie reviews, we learn a classifier to predict if a new movie review is positive or negative.

There are many ways to learn a text classifier. We only need to represent the corpus in a way that can be easily consumed by the classifier functions. The document term matrix is a good representation for this purpose as the rows are documents and columns are the words occurring in the corpus. This is very similar to data frames we used in previous chapters, where the data points are the rows and variables are the columns. Using a document term matrix, we can apply all classification algorithms towards learning a text classifier that we studied in Chap. 7.

8.6.2.1 Creating the Corpus and Document Term Matrix

To learn a document classifier, we need to have both positive and negative documents in the same corpus object. This is important because when we construct the document term matrix, we need the documents of both classes represented using the same set of words. If we have separate corpora for positive and negative documents, the respective document term matrices would only contain words occurring in that corpus. It is not convenient to obtain the global document term matrix from the two matrices.

We first obtain a source object for the base directory. By using the argument `recursive=T`, the source object contains the text documents in all the subdirectories.

```
> source = DirSource('review_polarity/',recursive=T)
```

The `source` object contains text files listed in an alphabetical order. We can see this by printing `source`. Due to this, the files containing negative reviews in the directory `review_polarity/txt_sentoken/neg/` are listed before the files containing positive reviews in the directory `review_polarity/txt_sentoken/pos/`. There are 1000 text files in each of these directories.

We obtain the corpus object from `source` below.

```
> corpus = Corpus(source)
```

The first 1000 text documents in `corpus` are negative reviews and the next 1000 text documents are positive reviews.

We obtain the document term matrix for the `corpus` object with the standard pre-processing options: removing stop words and punctuations, stemming, and applying TF-IDF weighting function.

```
> dtm = DocumentTermMatrix(corpus,
            control = list(weighting = weightTfIdf,
            stopwords = T, removePunctuation = T,
            stemming = T))
```

Most classification algorithms do not directly work with document term matrices. We first need to convert these into data frames. We first convert the document term matrix into a regular matrix and then into a data frame `x.df`.

```
> x.df = as.data.frame(as.matrix(dtm))
```

The variables or columns of the data frame are the words of the corpus, whereas the rows are the individual documents. We see this by printing the dimensions of `x.df`.

```
> dim(x.df)
[1]  2000 31255
```

We add an additional variable for the class label of the document: 1 and 0 representing positive and negative documents respectively.[5] As the first 1000 documents are negative reviews and the next 1000 documents are positive reviews, we create a vector of 'neg' repeated 1000 times followed by 'pos' repeated 1000 times.

```
> x.df$class_label = c(rep(0,1000),rep(1,1000))
```

8.6.2.2 Creating Train and Test Datasets

Similar to the examples in Chap. 7, we split our data `x.df` into train and test datasets so that we measure how good our classifier is. A simple way to randomly split a data frame is using the `split()` function that takes a variable with true/false values as input.

We construct an `index` variable containing a permutation of the row numbers of the data frame.

```
> index = sample(nrow(x.df))
```

[5] We could also use a factor variable with values "pos" and "neg." We are using 1 and 0 because it is convenient to train a logistic regression model with these class labels.

As `x.df` contains 2000 rows, `index` contains the elements in the sequence 1:2000 in random order.

```
> index
   [1] 1206 1282  576 1243 1063  386 1125 1222 ...
```

We split the data frame so that training dataset contains 1500 documents and test dataset contains 500 documents. For that we use the `split()` function with the argument `index > 500`.

```
> x.split = split(x.df, index > 1500)
```

The output `x.split` is a list of data frames containing the training and test datasets. We assign them respectively to the two data frames `train` and `test`.

```
> train = x.split[[1]]
> test = x.split[[2]]
```

We can check to see if the dimensions of these data frames are as expected.

```
> dim(train)
[1]  1500 31256
> dim(test)
[1]   500 31256
```

Our original data frame had 1000 positive and 1000 negative text documents. As we obtain `train` and `test` data frames by a random split, the distribution of class labels will not necessarily be equal. We use the `table()` function to see the distribution of positive and negative documents in `train`.

```
> table(train$class_label)

  0   1
760 740
```

We see that there are 760 negative documents and 740 positive documents in `train`. Not too far off from a 750-750 split. Similarly, we see that `test` contains the remaining 240 negative and 260 positive documents.

```
> table(test$class_label)

  0   1
240 260
```

8.6.2.3 Learning the Text Classifier

Once we have the data frames for the training data `train`, it is easy to learn the text classifier using any of the classification functions we looked at in Chap. 7. In this section, we train a text classifier using logistic regression and support vector machines (SVMs).

When learning any classification model over text data, one recurrent issue is data dimensionality which is equal to the number of words in the corpus. The `train` data frame is very high dimensional: it contains 31,256 variables, which means there are $1500 \times 31{,}255$ entries in the data frame. We can learn a classifier in R over such a large data frame, but it is fairly expensive in terms of computation time and memory.

A good approximation is to use a subset of features when learning the text classifier. It is an approximation because the classifier using subset of features might not be as good as the classifier learned using the complete set of variables. On the other hand, if we choose the variables in a smart way, this classifier is likely to have an accuracy close to the latter one.

Choosing variables in a classification model falls under the area of feature selection, which is a deep area of machine learning by itself. Here, we use a simple heuristic to select the variables: by using the frequent terms of the corpus. As we have already removed stop words, and are using TF-IDF weighting function, the frequently occuring terms in the corpus are the ones that are the most informative. We find the frequently occuring terms using the `findFreqTerms()` function.

```
> s = findFreqTerms(dtm,4)
> s
  [1] "act"          "action"        "actor"
  [4] "actual"       "alien"         "almost"
  ...
```

We empirically set a lower bound of 4 so that we end up with a reasonable number of variables. This factor is a trade-off between accuracy and computational cost.[6] By using this lower bound, we have 141 frequently occuring terms.

```
> length(s)
[1] 141
```

We add the variable `class_label` to `s` as we would need it to learn the text classifier.

```
> s = c(s,'class_label')
```

We obtain the data frame containing the variables as determined by `s` using `train[,s]`.

We first learn a logistic regression classifier using the `glm()` function. We specify `train[,s]` as the second argument.

```
> model.glm = glm(class_label ~ ., train[,s],
                  family='binomial')
```

Let us look at some of the model coefficients using the `coef()` function.

[6] In practice, the returns on using more number of variables are diminishing; after a point, we only see a marginal improvement in accuracy by adding more variables.

```
> coef(model.glm)
 (Intercept)           act         action         actor
  -0.2245104   -14.4608005     -7.6335646   -33.9385114

      actual         alien         almost          also
 -31.1075874    11.4464657     14.3002912   104.8285575

    although      american           anim         anoth
  43.0003199    42.7002215     10.4214973   -10.1282664

      around       audienc           back           bad
  -5.4040998   -25.4028306     38.8096867  -161.1490557

       becom         begin           best        better
 -10.9462855    -8.9060229     12.3624355   -38.3186012

         big       brother            can          cast
 -22.5387636   -10.1029090    -55.2868626     9.7951972

     charact          come         comedi           day
 -19.1775238   -12.8608680     13.9084740     0.3077193

      ...
```

We see that the word "bad" has a coefficient of -161.15, which is the lowest. According to our model, having one instance "bad" of the text would make the review negative. Similarly, the word "best" has a coefficient of 12.36. This is not true for all positive sentiment words as the word "better" has a negative coefficient of -38.32.

We see how well does this classifier do on our test dataset. We obtain the predictions of our model using the `predict()` function. We normalize these values to 0 and 1, which we then compare to the `class_label` variable of `test`.

```
> p = ifelse(predict(model.glm,test) < 0,0,1)
> table(p == test$class_label)/nrow(test)

FALSE  TRUE
0.268 0.732
```

We see that our model `model.glm` achieves 73.2 % accuracy.

As we discussed earlier, natural language is inherently ambiguous, which makes it hard to do text classification and text mining in general with high accuracy. It is interesting to look at some of the misclassified documents. We see that the model assigns a very low score of -4.15 to one of the positive reviews in the `test` dataset: `pos/cv107_24319.txt`. Using the sigmoid function (Sect. 7.2.2), we see that this score implies that the probability of the review being positive is 0.0155 or 1.55 %. Why does our model classify this document so incorrectly?

The answer to this question is clear when we see the contents of this document.

```
pos/cv107_24319.txt

is evil dead ii a bad movie?
it's full of terrible acting, pointless violence,
and plot holes yet it remains a cult classic nearly
fifteen years after its release.
```

The document contains the words "bad," "terrible," and "pointless," and then goes on to say that the movie is actually good. The presence of such negative sentiment words cause the classifier to label this document with a strong negative score, while the postive words are not sufficient to compensate for this.

Finally, we use support vector machines to learn the text classifier. We use the `svm()` function of the `e1071` package (Sect. 7.2.3). We reuse the `train[,s]` data frame containing the frequently occuring terms as variables. We use the Gaussian or radial kernel with cost = 2.

```
> model.svm = svm(class_label ~ ., train[,s],
                  kernel = 'radial', cost=2)
> model.svm

Call:
svm(formula = class_label ~ ., data = train[, s],
    kernel = "radial", cost = 2)


Parameters:
   SVM-Type:  eps-regression
 SVM-Kernel:  radial
       cost:  2
      gamma:  0.007142857
    epsilon:  0.1
Number of Support Vectors:  1410
```

We can see that the SVM classifier contains a large number of support vectors: 1410.

We also measure the accuracy of this text classifier on the `test` data.

```
> p.svm = ifelse(predict(model.svm,test) < 0.5,0,1)
> table(p.svm == test$class_label)/nrow(test)

FALSE  TRUE
0.256 0.744
```

We see that the SVM classifier has a slightly higher accuracy of 74.4 % as compared to that of the logistic regression classifier.

8.7 Chapter Summary

In this chapter we looked at analyzing text data with R using the `tm` package. When using this package, the first task is to load the text dataset in a Corpus object where each documents is represented by a TextDocument object. Before we begin analyzing the text dataset, it is important to perform certain preprocessing tasks to remove the noise from the data. The preprocessing tasks include: removal of punctuations, stopwords, and performing stemming to replace individual words or terms by their stem or basic form.

For most text analysis tasks, it is useful to represent the corpus as a term document matrix, where the individual terms of the documents are rows and documents are columns. The values stored in the matrix is given by its weighting function; the default weighting function is term frequently (TF) where we store the frequency of occurence of the term in the document. A more powerful weighting function is TF-IDF, which accounts for the importance of terms based on how infrequently the term occurs in the corpus.

The first text mining application we looked at was frequency analysis, where we find frequently occuring terms in a corpus as a glimpse into its overall sentiment. We also looked at finding frequently co-occuring terms for a given term to identify word associations. We also looked at text classification where we learn a classifier to label text documents. Using the term document matrix representation, we can apply all of the classification techniques that we have discussed in the previous chapters.

MIX
Papier aus verantwortungsvollen Quellen
Paper from responsible sources
FSC® C105338

If you have any concerns about our products,
you can contact us on
ProductSafety@springernature.com

In case Publisher is established outside the EU,
the EU authorized representative is:
Springer Nature Customer Service Center GmbH
Europaplatz 3, 69115 Heidelberg, Germany

Printed by Libri Plureos GmbH
in Hamburg, Germany